Artificial Intelligence in Engineering Design

VOLUME III

KNOWLEDGE ACQUISITION,
COMMERCIAL SYSTEMS,
AND INTEGRATED ENVIRONMENTS

Artificial Intelligence in Engineering Design

VOLUME III

KNOWLEDGE ACQUISITION,
COMMERCIAL SYSTEMS,
AND INTEGRATED ENVIRONMENTS

EDITED BY

CHRISTOPHER TONG
DEPARTMENT OF COMPUTER SCIENCE
RUTGERS UNIVERSITY
NEW BRUNSWICK, NEW JERSEY

DUVVURU SRIRAM
INTELLIGENT ENGINEERING SYSTEMS LABORATORY
MASSACHUSETTS INSTITUTE OF TECHNOLOGY
CAMBRIDGE, MASSACHUSETTS

ACADEMIC PRESS, INC.

Harcourt Brace Jovanovich, Publishers

Boston   San Diego   New York
London   Sydney   Tokyo   Toronto

ACADEMIC PRESS, INC.
1250 Sixth Avenue
San Diego, CA 92101-4311

United Kingdom Edition published by
ACADEMIC PRESS LIMITED
24-28 Oval Road, London NW1 7DX

ISBN 0-12-660563-7

Printed in the United States of America

92 93 94 95   BB   9 8 7 6 5 4 3 2 1

C. TONG DEDICATES THIS BOOK TO
SRI DA AVABHASA,
WITH LOVE AND GRATITUDE

D. SRIRAM DEDICATES THIS BOOK TO
THE GOOD LORD,
FOR CREATING THE HUMAN, WHO
CREATED THE COMPUTER,
WHICH HELPED GENERATE
THIS BOOK

Contents

Contributors .... *xiii*

Preface .... *xv*

1. **INTRODUCTION**
   C. Tong and D. Sriram **1**

PART VI:
KNOWLEDGE ACQUISITION

2. **A Knowledge Transfer Methodology in Selection**
   of Mechanical Design Components
   M. Waldron and K. Waldron **57**

PART VII:
COMMERCIAL APPLICATIONS

3. **ESKIMO: An Expert System for Kodak Injection Molding**
   R. Gammons **81**

4. **Product and Forming Sequence Design for Cold Forging**
   K. Sevenler **105**

5. **An Open-Architecture Approach to Knowledge-based CAD**
   D. Mishelevich, M. Katajamäki, T. Karras, A. Axworthy,
   H. Lehtimäki, A. Riitahuhta, and R. Levitt **125**

6. **Knowledge-Based Engineering Design at Xerox**
   L. Heatley and W. Spear **179**

vii

PART VIII:
INTEGRATED ENVIRONMENTS

7. An Intelligent CAD System for Mechanical Design
   N. Chao 199

8. The Expert Cost and Manufacturability Guide:
   A Customized Expert System
   P. London, B. Hankins, M. Sapossnek, and S. Luby 223

9. Engineous: A Unified Method for Design Automation, Optimization,
   and Integration
   S. Tong, D. Powell, and D. Cornett 235

10. A Unified Architecture for Design and Manufacturing Integration
    S. Kim 255

11. Dual Design Partners in an Incremental Redesign Environment
    M. Silvestri 281

12. DICE: An Object-Oriented Programming Environment for
    Cooperative Engineering Design
    D. Sriram, R. Logcher, N. Groleau, and J. Cherneff 303

PART IX:
THE STATE OF THE FIELD

13. Creating a Scientific Community at the Interface Between
    Engineering Design and AI: A Workshop Report
    D. Steier 367

Index 381

Contents of Volume I

1. INTRODUCTION
   C. Tong and D. Sriram

PART I: DESIGN REPRESENTATION

2. Representation and Control for the Evolution of VLSI Designs:
   An Object-Oriented Approach
   P. Subrahmanyam

3. Tools and Techniques for Conceptual Design
   D. Serrano and D. Gossard

4. Automobile Transmission Design as a Constraint Satisfaction
   Problem
   B. Nadel and J. Lin

5. Design Compilers and the Labelled Interval Calculus
   D. Boettner and A. Ward

6. Knowledge Representation for Design Creativity
   J. Hodges, M. Flowers, and M. Dyer

PART II: MODELS OF ROUTINE DESIGN

7. Investigating Routine Design Problem Solving
   D. Brown and B. Chandrasekaran

8. Design as Top-Down Refinement Plus Constraint Propagation
   L. Steinberg

9. A Knowledge-Based Framework for Design
   S. Mittal and A. Araya

10. BioSep Designer: A Process Synthesizer for Bioseparations
    C. Siletti and G. Stephanopoulos

11. VT: An Expert Elevator Designer That Uses Knowledge-Based
    Backtracking
    S. Marcus, J. Stout, and J. McDermott

12. A Design Process Model
    F. Brewer and D. Gajski

13. WRIGHT: A Constraint-Based Spatial Layout System
    C. Baykan and M. Fox

14. Designer: A Knowledge-Based Graphic Design Assistant
    L. Weitzman

Contents of Volume II

1. INTRODUCTION
   C. Tong and D. Sriram

PART III: MODELS OF INNOVATIVE DESIGN

2. Automated Reuse of Design Plans in BOGART
   J. Mostow, M. Barley, T. Weinrich

3. ARGO: An Analogical Reasoning System for Solving Design Problems
   M. Huhns and R. Acosta

4. Retrieval Strategies in a Case-Based Design System
   K. Sycara and D. Navinchandra

5. Case-Based Design: A Task Analysis
   A. Goel and B. Chandrasekaran

6. Function Sharing in Mechanical Design
   K. Ulrich and W. Seering

7. ALADIN: An Innovative Materials Design System
   M. Rychener, M. Farinacci, I. Hulthage, and M. Fox

8. The ADAM Design Planning Engine
   D. Knapp and A. Parker

9. Using Exploratory Design to Cope with Problem Complexity
   C. Tong

PART IV: REASONING ABOUT PHYSICAL SYSTEMS

10. Temporal Qualitative Analysis: Explaining How Physical Systems Work
    B. Williams

PART V: REASONING ABOUT GEOMETRY

11. Studies of Heuristic Knowledge-Based Approaches for Automated
    Configuration Generation and Innovation
    G. Nevill

12. A Case-Based Approach to the Design of Mechanical Linkages
    G. Kramer and H. Barrow

13. Mechanism Comparison and Classification for Design
    L. Joskowicz

14. How and Why To Get an AI or CAD/CAM System to Look at Your Objects
    L. Leff and D. Yun

Contributors

Numbers in parentheses refer to the pages on which the authors' contributions begin.

ALAN AXWORTHY (125), Design Power Inc., 10020 DeAnza Boulevard, Cupertino, California 95014.

NIEN-HUA CHAO (199), AT&T/Bell Laboratories, Murray Hill, New Jersey 07974.

J. CHERNEFF (303), Intelligent Engineering Systems Laboratory, Department of Civil Engineering, Massachusetts Institute of Technology, Cambridge, Massachusetts 02139.

DANNY CORNETT (235), General Electric Aircraft Engines, One Neumann Way, Evendale, Ohio 45215.

RICHARD A. GAMMONS (81), Computer Integrated Manufacturing Division, 7th Floor, Building 23, Eastman Kodak Company, 1669 Lake Avenue, Rochester, New York 14650.

N. GROLEAU (303), Intelligent Engineering Systems Laboratory, Department of Civil Engineering, Massachusetts Institute of Technology, Cambridge, Massachusetts 02139.

BLAIR HANKINS (223), Lotus Development Corporation, One Roger Street, Cambridge, Massachusetts 02139.

LYNN C. HEATLEY (179), Heatley Consulting Company, Suite 251, 3349 Monroe Avenue, Rochester, New York 14618-5591.

TAPIO KARRAS (125), Design Power Inc., 10020 DeAnza Boulevard, Cupertino, California 95014.

MATTI KATAJAMÄKI (125), Design Power Inc., 10020 DeAnza Boulevard, Cupertino, California 95014.

STEVEN H. KIM (255), Knowledge Systems Program, Laboratory for Manufacturing and Productivity, Massachusetts Institute of Technology, Cambridge, Massachusetts 02139.

HANNU LEHTIMÄKI (125), Design Power Inc., 10020 DeAnza Boulevard, Cupertino, California 95014.

RAYMOND E. LEVITT (125), Center for Facility Engineering, Stanford University, Stanford, California 94305.

R.D. LOGCHER (303), Intelligent Engineering Systems Laboratory, Department of Civil Engineering, Massachusetts Institute of Technology, Cambridge, Massachusetts 02139.

PHIL LONDON (223), Lotus Development Corporation, One Roger Street, Cambridge, Massachusetts 02139.

STEVEN LUBY (223), 346 Clyde Street, Brookline, Massachusetts 02167.

DAVID J. MISHELEVICH (125), Design Power Inc., 10020 DeAnza Boulevard, Cupertino, California 95014.

DAVID POWELL (235), Corporate Research and Development, General Electric Company, Post Office Box 8, Schenectady, New York 12301.

ASKO RIITAHUHTA (125), Tampere University of Technology, Laboratory of Machine Design, Post Office Box 527, SF 33101, Tampere, Finland.

MARK SAPOSSNEK (223), Cognition Inc., 900 Tech Park Drive, Billerica, Massachusetts 01821.

KORHAN SEVENLER (105), Battelle, 505 King Avenue, Columbus, Ohio 43201.

MARK J. SILVESTRI (281), Advanced Technology Consultant, 33 Malley Avenue, Avon, Massachusetts 02322.

WILLIAM J. SPEAR (179), Xerox Corporation, 8 Chelsea Way, Fairport, New York 14450.

DUVVURU SRIRAM (1, 303), Intelligent Engineering Systems Laboratory, Department of Civil Engineering, Massachusetts Institute of Technology, Cambridge, Massachusetts 02139.

DAVID STEIER (367), Engineering Design Research Center, Carnegie Mellon University, Pittsburgh, Pennsylvania 15213.

CHRISTOPHER TONG (1), Department of Computer Science, Rutgers University, New Brunswick, New Jersey 08903.

SIU SHING TONG (235), Corporate Research and Development, General Electric Company, Post Office Box 8, Schenectady, New York 12301.

KENNETH J. WALDRON (57), Department of Engineering Graphics, 2070 Neil Avenue, Ohio State University, Columbus, Ohio 43210-1358.

MANJULA B. WALDRON (57), Department of Engineering Graphics, 2070 Neil Avenue, Ohio State University, Columbus, Ohio 43210-1358.

Preface

The three-volume collection, "Artificial Intelligence in Engineering Design", has been put together incrementally over the course of the last six years. Most of the research efforts described herein are ongoing and thus chapters originally written early on in the enterprise are still representative of the state of the field. Some of these chapters additionally include updates that indicate the current status of the work.

For a variety of reasons, the order of the editors' names was chosen at random and fixed to be the same for each of the three volumes. However, both editors contributed equally to the making of all three volumes.

The editors would like to gratefully acknowledge the support and computational resources provided by the Computer Science Department of Rutgers University and the Intelligent Engineering Systems Laboratory at MIT, during the making of this collection.

Chapter 1
INTRODUCTION

Chris Tong and Duvvuru Sriram

1.1. WHAT THIS BOOK IS ABOUT

What is *design*? Design is the process of constructing a description of an artifact that satisfies a (possibly informal) functional specification, meets certain performance criteria and resource limitations, is realizable in a given target technology, and satisfies criteria such as simplicity, testability, manufacturability, reusability, etc.; the design process itself may also be subject to certain restrictions such as time, human power, cost, etc.

Design problems arise everywhere, and come in many varieties. Some are born spontaneously amidst the circumstances of ordinary human lives: design a dish for dinner that uses last night's leftovers; design some kind of hook-like artifact that will enable me to retrieve a small object that fell down a crack; design a "nice-looking" arrangement of the flowers in a vase. Other design problems are small but commercial in nature: design a paper clip-like device that doesn't leave a mark on the paper; design a lamp whose light can be turned to aim in any particular direction; design an artifact for storing up to twenty pens and pencils, in an easily accessible fashion. Still other design problems are formidable, and their solutions can require the efforts and coordination of hundreds of people: design a space shuttle; design a marketable electric car; design an international trade agreement; etc.

Because design is so ubiquitous, anything generic we can say about the *design process* -- the activities involved in actually solving a design problem -- can have great impact. Even better would be to provide active help to designers.

This book is all about how ideas and methods from Artificial Intelligence can help engineering designers. By "engineering design", we primarily mean the design of *physical artifacts* or *physical processes* of various kinds. In this book, we will see the design of a wide variety of artifacts exemplified, including: circuits and chips (Volume I, Chapters 2, 8, 12 and Volume II, 2, 8, 9, 10), swinging doors (Volume I, Chapter 6), copying machines (Volume I, Chapter 9 and Volume III, Chapter 6), cantilever beams (Volume I, Chapter 3), space telescopes (Volume II, Chapter 5), air cylinders (Volume I, Chapter 7), protein purifaction processes (Volume I, Chapter 10), fluid-mechanical devices (Volume II, Chapters 4 and 6), new alloys (Volume II, Chapter 7), graphics interfaces (Volume I, Chapter 14), automobile transmissions (Volume I, Chapter 4), spatial layouts (Volume I, Chapter 13), elevators (Volume I, Chapter 11), light-weight load-bearing structures (Volume II, Chapter 11), mechanical linkages (Volume II, Chapter 12), buildings (Volume III, Chapter 12), etc.

What you will not find in this book is anything on AI-assisted software design. On this point, our motivation is twofold: no book can (or should try to) cover everything; and AI and software engineering has already been treated in a number of edited collections (including [15, 30]).

This book is an edited collection of key papers from the field of AI and design. We have aimed at providing a state of the art description of the field that has coverage and depth. Thus, this book should be of use to engineering designers, design tool builders and marketeers, and researchers in AI and design. While a small number of other books have surveyed research on AI and design at a particular institution (e.g., [12, 31]), this book fills a hole in the existing literature because of its breadth.

The book is divided into three volumes, and a number of parts. This first chapter provides a conceptual framework that integrates a number of themes that run through all of the papers. It appears at the beginning of each of the three volumes. Volume I contains Parts I and II, Volume II contains Parts III, IV, and V, and Volume III contains Parts VI through IX.

Part I discusses issues arising in *representing* designs and design information. Parts II and III discuss a variety of models of the design process; Part II discusses models of routine design, while Part III discusses innovative design models. We felt that creative design models, such as they are in 1991, are still at too preliminary a stage to be included here. However, [11] contains an interesting collection of workshop papers on this subject. Parts IV and V talk about the formalization of common sense knowledge (in engineering) that is useful in many design tasks, and the reasoning techniques that accompany this knowledge; Part IV discusses knowledge about physical systems, while Part V gives several examples of formalized geometry knowledge. Part VI discusses techniques for acquiring knowledge to extend or improve a knowledge-based system. Part VII touches on the issue of building a knowledge-based design system; in particular, it presents a number of commercially available tools that may serve as modules within a larger knowledge-based system. Part VIII contains several articles on integrating design with the larger engineering process of which it is a part; in particular, some articles focus on designing for manufacturability. Finally, Part IX contains a report on a workshop in which leaders of the field discussed the state of the art in AI and Design.

1.2. WHAT DOES AI HAVE TO OFFER TO ENGINEERING DESIGN?

In order to answer this question, we will first examine the nature of engineering design a little more formally. Then we will briefly summarize some of the major results in AI by viewing AI as a software engineering methodology. Next we will look at what non-AI computer assistance is currently available, and thus what gaps are left that represent opportunities for AI technology. Finally, we outline how the AI software engineering methodology can be applied to the construction of knowledge-based design tools.

1.2.1. Engineering Design: Product and Process

Engineering design involves mapping a specified *function* onto a (description of a) realizable physical *structure* -- the designed artifact. The desired function of the artifact is what it is supposed to do. The artifact's physical structure is the actual physical parts out of which it is made, and the part-whole relationships among them. In order to be realizable, the described physical structure must be capable of being assembled or fabricated. Due to restrictions on the available assembly or fabrication process, the physical structure of the artifact is often required to be expressed in some *target technology*, which delimits the kinds of parts from which it is built. A *correct design* is one whose physical structure correctly implements the specified function.

Why is design usually not a classification task [6], that is, a matter of simply looking up the right structure for a given function in (say) a parts catalog? The main reason is that the mapping between function and structure is not simple. For one thing, the connection between the function and the structure of an artifact may be an indirect one, that involves determining specified behavior (from the specified function), determining actual behavior (of the physical structure), and ensuring that these match. For another, specified functions are often very complex and must be realized using complex organizations of a large number of physical parts; these organizations often are not hierarchical, for the sake of design quality. Finally, additional non-functional constraints or criteria further complicate matters. We will now elaborate on these complications.

Some kinds of artifacts -- for example, napkin holders, coat hangers, and bookcases -- are relatively "inactive" in the sense that nothing is "moving" inside them. In contrast, the design of a *physical system* involves additionally reasoning about the artifact's *behavior*, both external and internal. The external behavior of a system is what it does from the viewpoint of an outside observer. Thus, an (analog) clock has hands that turn regularly. The internal behavior is based on observing what the *parts* of the system do. Thus, in a clock, we may see gears turning. Behavior need not be so visible: electrical flow, heat transmission, or force transfer are also forms of behavior.

In a physical system, behavior *mediates* function and structure. The *function* is achieved by the *structure behaving* in a certain way. If we just possessed the physical structure of a clock, but had no idea of how it (or its parts) behaved, we would have no way of telling that it achieves the function of telling time.

Not only in a physical system but also in *designing* a physical system, behavior tends to act as intermediary between function and structure. Associated with a specified function is a *specified behavior*; we would be able to tell time if the angle of some physical structure changed in such a way that it was a function of the time. Associated with a physical structure is its *producible behavior*; for example, a gear will *turn*, provided that some rotational force is applied to it. In rough terms then, designing a physical system involves selecting (or refining) a physical structure (or description thereof) in such a way that its producible behavior matches the specified behavior, and thus achieves the desired function. Thus, we could successfully refine the "physical structure whose angle is a function of the hour" as either the hand on an electromechanical clock, or as the shadow cast by a sundial.

Complex functions often require complex implementations. For example, a jet aircraft consists of thousands of physical parts. Parts may *interact* in various ways. Thus the problems of *designing* the parts also interact, which complicates the design process. Such interactions (among physical parts or among the problems of designing those parts) can be classified according to their strength.

For instance, many parts of the aircraft (the wings, the engine, the body, etc.) must, together, behave in such a way that the plane stays airborne; thus the subproblems of designing these parts can be said to *strongly interact* with respect to this specification of global behavior. Non-functional requirements such as global resource limitations or optimization criteria are another source of strong interactions. For example, the total cost of the airplane may have to meet some budget. Or the specification may call for the rate of fuel consumption of the plane to be "fairly low". Not all ways of implementing some function may be equally "good" with respect to some global criterion. The design process must have some means for picking the best (or at least a relatively good) implementation alternative. Good implementations often involve *structure-sharing*, i.e., the mapping of several functions onto the same structure. For example, the part of the phone which we pick up serves multiple functions: we speak to the other person through it; we hear the other person through it; and it breaks the phone connection when placed in the cradle. Important resources such as "total amount of space or area" and "total cost" tend to used more economically through such structure-sharing. On the other side of the coin, allowing structure-sharing complicates both the representation of designs and the process of design.

That neighboring parts must fit together -- both structurally and behaviorally

-- exemplifies a kind of *weak* or *local interaction*. Thus the wings of the plane must be attachable to the body; the required rate of fuel into the engine on the left wing had better match the outgoing rate of fuel from the pump; and so forth. The *design process* must be capable of ensuring that such constraints are met.

1.2.2. Artificial Intelligence as a Software Engineering Methodology

Now that we've briefly examined engineering design, we will equally briefly examine (the most relevant aspects of) Artificial Intelligence (AI).

Problem-solving as search. The late 1950s and the 1960s saw the development of the *search paradigm* within the field of Artificial Intelligence. Books such as "Computers and Thought" [10], which appeared in 1963, were full of descriptions of various weak methods whose power lay in being able to view the solving of a particular kind of problem as search of a space. In the late 1960s, the notion of heuristic search was developed, to account for the need to search large spaces effectively.

Knowledge as power. Nonetheless, most of the problems considered in those early days were what are now commonly called "toy problems". As the 1970s began, many practitioners in the field were concerned that the weak methods, though *general*, would never be *powerful* enough to solve real problems (e.g., medical diagnosis or computer configuration) effectively; the search spaces would just be too large. Their main criticisms of the earlier work were that solving the toy examples required relatively little knowledge about the domain, and that the weak methods required knowledge to be used in very restrictive and often very weak ways. (For example, in state space search, if knowledge about the domain is to be used, it must be expressed as either operators or evaluation functions, or else in the choice of the state space representation.) Solving real problems requires extensive knowledge. The "weak method" critics took an engineering approach, being primarily concerned with *acquiring* all the relevant knowledge and *engineering* it into some usable form. Less emphasis was placed on conforming the final program to fit some general problem-solving schema (e.g., heuristic search); more concern was placed on just getting a system that worked, and moreover, that would produce (measurably) "expert level" results. Thus was born the "expert systems" paradigm.

Evolution of new programming paradigms. Several list-processing languages were developed in the late 1950s and early 1960s, most notably, LISP. The simple correspondence between searching a space for an acceptable solution and picking an appropriate item in a list made the marriage of AI (as it was at that time) and list-processing languages a natural one. Various dialects of LISP evolved, and the developers of the main dialects began evolving programming environments whose features made LISP programming more user-friendly (e.g., procedural enrichments of a language that was originally purely functional; debuggers; file packages; windows, menus, and list structure editors).

At the same time as the "expert systems" paradigm was developing, a new wave of programming languages (often referred to as "AI languages") was arriving. Like the evolution of expert systems, this development phase seemed to be motivated by the need for less general (but more powerful) languages than LISP. Many of these languages were (part of) Ph.D. theses (e.g., MICROPLANNER [42, 47] and Guy Steele's constraint language [35]). Often these languages were built on top of the LISP language, a possibility greatly facilitated because of the way LISP uniformly represents both data and process as lists. Often these languages were never used in anything but the Ph.D. dissertation for which they were developed, because they were overly specialized or they were not portable.

Exploring tradeoffs in generality and power. During the 1970s, at the same time as many researchers were swinging to the "power" end of the "generality-power" tradeoff curve in their explorations, others were striking a middle ground. Some researchers, realizing the limitations of the weak methods, began enriching the set of general building blocks out of which search algorithms could be configured. New component types included: constraint reasoning subsystems, belief revision subsystems, libraries or knowledge bases of various kinds; a variety of strategies for controlling problem-solving, etc. Other programming language designers than those mentioned previously developed new, specialized (but not overly specialized), and portable programming paradigms, including logic programming languages, frame-based and object-oriented languages, and rule-based languages. Rule-based languages such as OPS5 arrived on the scene at an opportune moment. In many cases, their marriage to "expert systems" seemed to work well, because the knowledge acquired from observing the behavior of domain experts often took the simple associational (stimulus-response) form: "IF the problem is of type P, then the solution is of type S."

Synthesis, consolidation and formalization. AI researchers of the late 1950s and the 1960s posed the *thesis*, "Generality is sufficient for problem-solving." 1970s researchers criticized this thesis, claiming the resulting methods were insufficient for solving real problems, and responded with the *antithesis*, "Power is sufficient." However, that antithesis has been critiqued in turn: "Expert systems are too brittle"; "special languages only work for the application for which they were originally developed"; etc.

Since the early 1980s, AI seems to be in a period of *synthesis*. One useful tool for illustrating the kind of synthetic framework that seems to be emerging out of the last few decades of research is depicted in Figure 1-1. Rather than pitting generality against power, or the "search paradigm" against the "expert systems" or "knowledge-based paradigm", the framework unifies by providing three different levels of abstraction for viewing the same "knowledge-based system": the knowledge level; the algorithm level; and the program level.

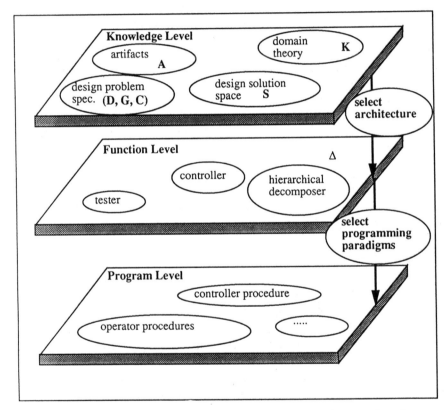

Figure 1-1: Rationally Reconstructed Knowledge-Based System Development

These three levels directly reflect the history of AI as we have just rationally reconstructed it. The "knowledge level" view of a knowledge-based system describes the knowledge that is used by and embedded in that system. The "algorithm level" view describes the system as a search algorithm, configured out of standard component types (e.g., generators, testers, patchers, constraint propagators, belief revisers, etc.). Finally the "program level" view expresses the system in terms of the elements of existing programming paradigms (rules, objects, procedures, etc.). Within the "algorithm level", a spectrum of search algorithms -- ranging from weak to strong methods -- can be created depending on the choice of component configuration, and the choice of how knowledge (at the knowledge level) is mapped into the search algorithm components. A similar set of choices exists relative to the mapping of the "algorithm level" search algorithms into "program level" knowledge-based systems.

Many of the ideas and insights of this period of synthesis can be viewed as either: stressing the importance of distinguishing these levels (e.g., [6]); introducing criteria for evaluating systems at the different levels (e.g., epistemological adequacy [17] at the knowledge level; (qualitative) heuristic adequacy [17] at the algorithm level; and (quantitative) heuristic adequacy at the program level); fleshing out the primitives at each level (e.g., ATMSs [7] or constraint propagators [36] at the algorithm level); understanding and validating established correspondences between entities at different levels (e.g., between search algorithms and list-processing languages; or expert knowledge and rule-based languages), or on discovering new correspondences.

AI as a software engineering methodology. Viewed as a software engineering methodology, AI works best for developing those knowledge-based systems whose construction is usefully aided by creating explicit knowledge level and function level abstractions. More specifically, the AI methodology works well when:

- the problems addressed by the desired knowledge-based system are ill-structured, and involve large or diverse types of knowledge (when expressed at the knowledge level);

- that knowledge can be incorporated into an *efficient* search algorithm, that can be viewed as a configuration of standard building blocks for search algorithms;

- that search algorithm, in turn, can be implemented as an *efficient* program, using currently available programming paradigms.

1.2.3. Computer-aided Design

1.2.3.1. Opportunities for AI in computer-aided design

In many design areas (e.g., VLSI design or mechanical design), progress in automating the design process passes through a sequence (or partial ordering) of somewhat predictable stages (see Table 1-1). As we see it, design tool developers proceed through the following stages: permitting design capture; automating specific expert tasks; constructing unifying representations and system architectures; modeling and automating the complete design process; automatically controlling the design process; automatically re-using design experience; automatically improving tool performance. The central intuition is that, with the passage of time, design tools play an increasingly more *active* role in the design process. Note that the sequence is not meant to imply that the user is (or should ever be) removed from the design process; instead, the user receives increasingly greater assistance (and a more cooperative and powerful design environment) with time. Table 1-2 lists some particular technological contributions that have been made to design automation by academia and by industry.

Permitting design capture. In the beginning, graphical editors are created that allow the user to enter, visualize, modify, and store new designs, as well as retrieve old designs, in a simple manner. This is such a universal starting point for design automation in any new area that "CAD/CAM" (Computer-Aided Design/Computer-Aided Manufacturing) tends to be used as a synonym for fancy graphical, object-oriented interfaces. The development of these tools is largely aided by techniques drawn from graphics and database management (including such AI-related areas as deductive or object-oriented databases).

Automating the execution of expert tasks. As time passes, tool users become less satisfied with a purely passive use of CAD. CAD tool builders identify specific *analysis* and *synthesis* tasks which have been carefully delimited so as to be automatically executable (e.g., placement, routing, simulation). AI research can make a contribution at this stage; the software engineering methodology mentioned in Section 1.2.2 can facilitate the incremental creation, testing, and development of knowledge-based systems which carry out the more ill-structured analysis and synthesis tasks. (Well-structured tasks are carried out by algorithms.)

Constructing unifying representations and system architectures. A problem of integration arises; the set of available CAD tools is growing increasingly richer, but the tools are diverse, as are the design representation languages they

Table 1-1: Stages in the Evolution of Design Automation

DESIGN AUTOMATION GOAL	PROBLEM	AI ISSUE
Permit design capture	What functions does the user interface provide?	Deductive or object-oriented databases
Build tools for specific tasks	How to automate specialized types of reasoning?	Inference; Expert systems
Integrate tools	How to communicate between tools?	Representation; Architectures
Manage versions	Which task, tool, parameters?	Search space
Model design process	Which model is right for the task?	Taxonomy of tasks and corresponding methods
Find good design fast	How to guide choices?	Control
Improve design system	Where and how to improve?	Machine learning
Reuse design knowledge	How to acquire? How to re-use?	Machine learning, Case-based reasoning

utilize. AI can enter again to contribute ideas about unifying representation languages (e.g., object-oriented languages) that enable the creation of "design toolboxes", and unifying system architectures (e.g., blackboard architectures).

Modeling the design process. Having a single unified environment is good but not sufficient. How can we guarantee that we are making the most of our available tools? AI contributes the notion of the design process as a search through a space of alternative designs; the synthesis tools are used to help generate new points in this space; the analysis tools are used to evaluate the consistency, correctness, and quality of these points; the idea of search is used to guarantee that *systematic progress* is made in the use and re-use of the tools to generate new designs or design versions.

Table 1-2: Increasingly more Sophisticated Technological Contributions
From Industry and Academia

Technology	University	Industry	Design Automation Goal
Interactive graphics	Sketchpad (MIT, 1963)	DAC-1 (GM, early 60s)	design capture
Drafting (2D)		Autocad^{TM} ADE^{TM}	design capture
Solid modelers (3D) (CSG, BREP)	BUILD (UK) PADL (Rochester) (see [29]) MicroStation^{TM}	I-IDEAS^{TM} ACIS^{TM} + specific tools etc.	design capture
Solid modelers (super-quadrics, nonmanifold)	ThingWorld [28] Noodles (CMU)		design capture
Physical modelers (spatial + physics)	ThingWorld		design capture + specific tools
Parametric modelers (variational geometry + constraint management)	Work at MIT-CAD Lab PADL-2 (U. Rochester)\	DesignView^{TM} (2D) ICONEX^{TM} (2D) PRO/ENGINEER^{TM} (3D) Vellum^{TM}	design capture + specific tools
Semantic modeling + geometry (mostly wire frame) + constraint management + layout		ICAD WISDOM DESIGN++	design capture + specific tools
Logic synthesis (ECAD) [18, 27]		Logic Synthesizer^{TM}	design process model (algorithmic)
Concept generators (routine design)	VEXED DSPL CONGEN	PRIDE (in-house)	design process model
Concept generators (innovative design)	BOGART CADET EDISON KRITIK ALADIN DONTE etc.	ARGO (in-house)	design process model + control
Integrated frameworks (cooperative product development [33])	DICE (MIT, WVU) NEXT-CUT (Stanford)	integrate tools, version IBDE (CMU)	management

Controlling the design process. The priced paid for search is efficiency, as the search space is generally quite large. Exhaustive search of the space is usually intractable; however, a search which focuses its attention on restricted but promising *subspaces* of the complete design space may trade away the guarantee of an optimal solution (provided by exhaustive search), in return for an exponential decrease in overall design time.

How can a knowledge-based system control its search of a large design space so that a satisfactory solution is produced in a reasonable amount of time? Good *control heuristics* help.

Control heuristics may either be domain-specific or domain-independent. "Spend much of the available design time optimizing the component that is a bottleneck with respect to the most tightly restricted resource" is an example of a domain-independent heuristic, while "Spend much of the available design time optimizing the datapath" is a domain-specific version of this heuristic that applies to certain situations in the microprocessor design domain. Control heuristics may address different control questions. Some address the question: "Which area of the design should be worked on next?" while others address the question, "What should I do there? How should I refine that area of the design?"

Automatically improving performance and automated reuse of design experience. At this stage in the evolution of design automation in a design area, most of the burden of routine design has been lifted from the end user; this has been accomplished by *reformulating* this burden as one for the knowledge engineers and system programmers. In turn, techniques from machine learning can make life easier for the system builders themselves. In particular, they can build a design tool that is *incomplete* or *inefficient*; the design tool can be augmented by machine learning and case-based reasoning techniques that can extend the knowledge in the system, or use its knowledge with ever greater efficacy.

1.2.3.2. The differing goals of CAD tool and AI researchers

A misunderstanding frequently arises between AI researchers who develop experimental Computer-aided Design (CAD) tools, and traditional CAD tool developers in a particular design area (e.g., VLSI or mechanical design) who specialize in developing new design tools that will be usable in production mode in the near-term future. The CAD tool developers accuse the AI researchers of being too general, and of creating inefficient or toy knowledge-based systems. On the other hand, the AI researchers criticize the traditional CAD tool researchers of creating overly brittle systems.

Confusion arises because these two types of researchers (each of whom is likely to be reading this book) do not share quite the same research goals, and each tends to judge the other with respect to their own community's values. Traditional CAD tool developers seek to reduce the effort in creating *new designs*. Most AI researchers aim at reducing the effort in developing *new design tools*.

Both research programs are worthy enterprises. The former goal requires the design tools to be powerful. The latter requires the methodology for constructing the tool (e.g., instantiation of a particular shell) to be general, and thus sometimes requires the design tool itself to be an instance of a general form rather than a custom-built tool. This book describes results from both enterprises.

1.2.4. A Methodology for Building a Knowledge-based Design Tool

In Section 1.2.1, we described the problem of design, and mentioned features of the problem that indicate design is generally an *ill-structured problem*. We then described AI as a three-level, software engineering methodology for developing knowledge-based systems for solving ill-structured problems. In the last section, we identified specific design automation tasks where such a methodology can be most usefully applied. We now describe what the general methodology looks like, when restricted to building knowledge-based design systems.

The steps involved in the development of AI tools for engineering design are shown in Table 1-3. The rest of this chapter will go into these steps in greater detail. We indicate which levels are involved in each step (knowledge, function, or program level), and which sections of this chapter will elaborate on that step.

The next few sections flesh out basic ideas relevant to understanding the phases of this methodology. They also relate the later chapters of this book to this methodology.

1.3. FORMALIZING THE DESIGN SPACE AND THE DESIGN KNOWLEDGE BASE

Algorithms can be decomposed into *passive* data structures and *active* access and update operations on these data structures. Similarly, models of design can be partitioned into passive components -- the design space and the knowledge base; and an active component -- the process that (efficiently) navigates through that space, guided by the knowledge in the knowledge base. This section

Table 1-3: Phases of Knowledge-based Tool Construction

PHASE	LEVEL	SECTION
Identify design task	knowledge level	1.5.1
Formalize design space	algorithm level	1.3
Formalize knowledge base	algorithm level	1.3
Configure appropriate model of design process, based on properties of design task and design space	algorithm level, knowledge level	1.4, 1.5.2
Instantiate by acquiring and operationalizing knowledge	knowledge level, algorithm level	1.5.2
Implement	algorithm level, program level	1.5.3
Test (validate and verify)	all levels	covered in individual chapters
Deploy		covered in individual chapters
Improve	all levels	covered in individual chapters

focuses on the nature and organization of design spaces and design knowledge bases, while the next section explores the spectrum of design processes that search such a space.

1.3.1. What Distinguishes a Design Search Space?

In order to characterize a (dynamically generated) search space, we must define the nature of the points in that space, the relationships that can exist between points in the space, and how to generate new points from old ones.

Points in the design space. In a *design space*, the points can be design specifications[1] or implementations. They can be at varying levels of abstraction. Some points may only correspond to parts of a design (specification or implementation). A single such point P1 might have associated with it:

- its parts: {P11,...,P1n}. In the simplest case, these parts are simple parameters; in general, they can be arbitrarily complex structures.
- constraints on it and its parts.
- information about how its parts are connected.

Chapter 3 in Volume I considers the case where a design can be represented as a *constraint graph*, whose nodes are parameters, and whose arcs represent constraint relationships. Several design operations are easy to implement (in a domain-independent manner), given such a representation: automatic generation of parameter dependencies; evaluation of a constraint network; detection of over- and under-constrained systems of constraints, and the identification and correction of redundant and conflicting constraints. A few commercial tools, such as Cognition's MCAE^{TM} and DesignView^{TM} (see Volume III, Section 4.3.1), incorporate variations of Serrano's constraint management system. Chapter 4 in Volume I goes on to discuss how such a constraint network representation can be used to design automobile transmissions. The application of interval calculus methods to constraint satisfaction problems is treated in Volume I, Chapter 5. These interval methods are used in a mechanical design compiler, which accepts generic schematics and specifications for a wide variety of designs, and returns catalog numbers for optimal implementations of the schematics.

The design space as a whole. Some of the most basic relationships that can exist between points in the design space include:

- P2 is a part of P1.
- P2 is a refinement of P1 (where P1 and P2 are both specifications). P2 consequently may be at a lower level of abstraction than P1.
- P2 is an implementation of P1 (where P1 is a specification for and P2 is a description of an artifact in the target technology).

---

[1]We use the word *specification* to denote a *function* or a *goal* that needs to be realized or satisficed in the final design, e.g., "Design a land vehicle capable of going at least 100 mph over sand."

- P2 is an optimization of P1 (i.e., P2 is better than P1 with respect to some evaluatable criterion).

- P2 is a patch of P1 (i.e., P1 contains a constraint violation that P2 does not).

These points can also be clustered into multiple levels of abstraction; for example, in VLSI design, there might be a system level, a logic level, and a geometric layout level. Figure 1-2 illustrates some of these relationships.

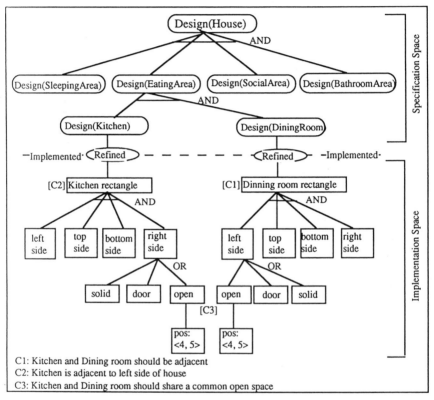

Figure 1-2: The Design Space as an AND/OR Tree

Dynamically generating the design space. Some of the most basic operations for generating new points in the design space from old ones include:

- refining P1 into P2.
- implementing P1 as P2 in target technology T.
- decomposing P1 into {P11,...,P1n}.
- optimizing P1 into P2 with respect to criteria O.
- patching constraint violation V in P1, yielding P2.

Chapter 2 in Volume I discusses the issues involved in representing all these aspects of a design space. The points are illustrated in the context of VLSI design.

1.3.2. What Distinguishes a Design Knowledge Base?

Often the parts that occur in designs (at any level of abstraction) can be viewed as *instances* of a generic class. For example, microprocessors are usually composed of generic parts such as ALUs, registers, busses, etc.

Such regularity can be exploited by maintaining a knowledge base of *design object classes*, and then viewing designs as configurations of instances of particular classes (e.g., a new microprocessor instance is constructed by creating an instance of ALU5, Datapath3, Bus4, etc. and then connecting these object instances together in a particular fashion). Design objects are also often *parameterized*. A complete instance of such a parameterized object class is created by assigning values to all the parameters.

In the standard object-oriented fashion, such design object classes may be organized hierarchically, thus reaping the standard benefits of inheritance. Design process operations (such as refinement, optimization, etc.) may also be indexed in a class-specific fashion (as methods), and thus, may also be inheritable.

The relation between a design space, a design knowledge base (of the kind just described), and a design process is as follows. A *design process* operation such as refinement, patching, or optimization may generate a new point in the *design space* from one or more old ones; the operation itself may involve creating new instances of design object classes from the *design knowledge base*.

Based on such an object-oriented representation of a design knowledge base, Chapter 2 (Volume I) discusses how to represent parameterized designs, design histories, and task-specific experts. As examples of desirable properties for design representations, it suggests modularity, compactness, flexibility permitted in the design process (e.g. in allowing both top-down and bottom-up design, and concurrent execution of design tasks), and extensibility; it describes how these properties may be achieved.

How does the design process know which design object class(es) should be instantiated to carry out a particular design operation (e.g., refinement of part P1)? One answer is to hardcode the association. For example, a specific *refinement rule* might express the knowledge that whenever a part of type P1 is being refined, it should be decomposed into parts of type {P11,...,P1n}. Or a specific *patching rule* might fix a specific type of constraint violation that commonly occurs in a specific kind of object. The design process models in Part II of this book take this hardcoded approach.

Another answer is to treat this question as a problem that must be solved explicitly by the design process. For example, the process of patching a constraint violation might actually involve solving the problem of *recognizing* that a particular object in the design is an instance of (or similar to) some object in the knowledge base, and then *recognizing* that the specified function of that object has been disabled in some way (by the context of the object). Available patching methods associated with that object class can then be applied (or adapted). Chapter 6 (Volume I) discusses how to organize a design knowledge base so that this kind of "innovative" patching can occur.

1.4. MODELS OF THE DESIGN PROCESS

1.4.1. The Nature of Design Tasks

1.4.1.1. Design task dimensions

Design tasks can be classified along several dimensions, including:

- available methods and knowledge;
- amount of unspecified (physical) structure;
- gap in abstraction levels between specification and implementation;
- complexity of interactions between subproblems; and
- amount and type of knowledge a system user can provide.

Available methods and knowledge. Is an appropriate method and/or sufficient knowledge always available for choosing what task to address next in the design process (e.g., what part to refine, what bug to fix, etc.)? Is knowledge or a method available for executing that next task? If there is more than one way of executing that task, is knowledge or a method available for selecting the alternative that will have the (globally) best outcome? The more (appropriate) knowledge and methods are available, the more *routine* the design task is. We will focus our discussion on two basic types of knowledge and methods: *generative* knowledge and methods, for generating new points in the design space; and *control* knowledge and methods, for helping the design process to converge efficiently on an acceptable design solution.

If sufficient knowledge and methods are available for always *directly* (i.e., without problem-solving) generating the next point in the design space and for converging on an acceptable design with little or no search, we will call the task a *routine* design task.

If the available knowledge and methods do allow for fairly rapid generation of an acceptable solution, but only by:

- *indirect* generation of new points in the design space -- i.e., finding a way to generate the next point in the design space involves a problem-solving process; and/or

- *indirect* control of the search, i.e., via problem-solving.

that is -- by itself, the available (directly applicable) knowledge generates unacceptable designs -- we will call the task an *innovative* design task.

Finally, if a problem-solving process is required to construct the design space in the first place, or if the best method available (given our current understanding) is an unguided search through a very large space, we will call the task a *creative* design task.

We will call design process models capable of handling these three types of design tasks routine, innovative, and creative design process models, respectively. We discuss routine design processes in Section 1.4.2, and innovative design processes in Section 1.4.3. We feel that creative design models, such as they are, are still at too preliminary a stage to be included here. However, [11] contains an interesting collection of workshop papers on this subject. Since we have tied creative design to the creation of the proper design space, creative design can also be viewed as a search through a space of design space representations, and thus work on problem reformulation and representation design can be seen as relevant here (see, e.g., [1]).

The terms "routine", "innovative", and "creative design" were introduced in [3], but were used in a somewhat different sense. Note that we use these terms in reference to the *task* and the *process*, but not the *product*. Thus, an innovative design process (e.g., replay of design plans) might not necessarily produce a product that is innovative with respect to the current market.

Amount of unspecified structure. Design maps function into (physical) structure. A design task often provides part of the (physical) structure of the design.

Since the design process involves creating a complete (physical) structure, it is also useful to identify what of the physical structure is left to be determined as a measure of design task complexity [39]. Design tasks are usefully distinguished according to what the *unspecified* structure looks like [40].

In *structure synthesis tasks*, the unspecified structure could potentially be any composition of primitive parts, which may not exist in the knowledge/data base. For example, the specified function might be a boolean function such as (and (or x y) (not z)). The physical structure might be any gate network that implements the boolean function; no part of the gate network is given *a priori*.

In *structure configuration tasks*, the unspecified structure is a configuration of parts of pre-determined type, and connectors of pre-determined type. For example, the physical structure might be a house floorplan containing some number of rooms, that can be connected by doors. For a particular floorplanning problem, the number of rooms and the size of the house might be given. In this case, the unspecified structure would be the configuration of rooms and doors, plus the values for room and door parameters.

Finally, in *parameter instantiation tasks*, the unspecified structure is the set of values for the parameters of each part. For example, the physical structure might be the part decomposition for all air cylinders (Volume I, Chapter 7). For a particular air cylinder design problem, the values for particular parameters (e.g., the length of the cylinder) might be given. Then the unspecified structure would be the values for all the remaining parameters.

Gap in abstraction levels between specification and implementation. In the simplest case, the design specification and the design implementation are at the same level of abstraction. This occurs, for example, when the only unspecified structure is parameter values. In other cases, a single level separates the functional specification from the target implementation level. That is, knowledge and methods are available for directly mapping the pieces of the specification into implementations; implementing a boolean function as a gate network is a simple example. In the worst case, the design may have to be driven down through several levels of abstraction before it is completed. For instance, in VLSI design, the initial specification might be of a digital system (e.g., a calculator or a microprocessor), which is first refined into a "logic level" description (a gate network), and then into a "layout level" description (of the actual geometry of the chip).

Complexity of interactions between subproblems. On one extreme (independent subproblems), the subproblems can all be solved independently, the solutions can be composed easily, resulting in an acceptable global design. On the other extreme, the subproblems strongly interact: a special (relatively rare) combination of solutions to the subproblems is required, and combining these solutions into an acceptable global solution may not be easy or quick. Complexity increases when the number of interactions increases or when the type of interaction becomes more complex.

Two major types of design interactions are worth distinguishing. *Compositional interactions* arise when not all choice combinations (for refining or implementing the different parts of the design) are (syntactically) composable. For example, in VLSI design, the output of one part may be "serial", while the input of another may be "parallel"; if the output of the one must feed the input of the other, then the parts are not syntactically composable. Syntactic interactions may be further subdivided into *functional interactions* index(Functional interactions) among parts of a functional decomposition (e.g., in VLSI design, the "serial output/input" interaction) and *physical interactions* among parts of the implementation (e.g., in VLSI design, wire1 and wire3 on the chip must be at least 3 lambda units apart).

Resource interactions arise when different choice combinations lead to different overall usage of one or more global resources (e.g., delay time, power, or area in VLSI design). Different resources "compose" in different ways: e.g., global part counts are related to local part counts by simple addition; global delay time involves critical path analysis; etc.

Each interaction can be represented by a *constraint*. A *local* constraint only constrains a single part; a *semi-local* constraint constrains a relatively small number of parts; and a *global* constraint constrains a relatively large number of parts. Compositional interactions tend to be represented by semi-local constraints (because the syntax rules for correctly composing parts tend to refer to a small number of parts). Resource interactions tend to be represented by global constraints (since the global resource usage tends to be a function of the whole design).

Compositional interactions are typically *weak* interactions; they are usually representable by semi-local constraints. In contrast, resource interactions are typically *strong* interactions, representable by global constraints.

Amount and type of knowledge a system user can provide. In considering the nature of a design task, we will consider human users as knowledge sources, and thus classify the design tasks addressed by a particular knowledge-based design system as "routine" or "innovative" depending on how much knowledge (and method) the system and the user *together* can provide during the overall design process. Thus, even if the design system itself has no directly applicable control knowledge, if the user makes choices at every decision point in a manner that leads to rapid convergence on an acceptable solution, then the task is "routine".

1.4.1.2. Design task decomposition

While sometimes the terms we have just introduced are appropriately applied to the design task as a whole, it is often the case that "the design task" is a collection of (themselves sometimes decomposable) subtasks. Whether a task is considered a "routine design task" really depends on whether the subtasks are all routine and on how strongly the subtasks interact; the same design task may have relatively more and less routine parts to it. A category such as "parameter instantiation task" may be aptly applied to one subtask, and be inappropriate for another. Reference [5] makes some further points about task decomposition and associating different methods with different types of subtasks.

1.4.2. Models of Routine Design

1.4.2.1. Conventional routine design

In many cases, knowledge-based models of design are simply inappropriate, or would constitute overkill; conventional methods suffice for solving the task (or subtask). Some design tasks can be cast as a set of linear constraints $C(s)$ on a set of real-valued variables, plus an objective function $O(s)$ on these variables; for such problems, the methods of *linear programming* apply. Other simple design tasks can be cast as *constraint satisfaction problems* (CSPs) when: only parameter values are left unspecified; each parameter has a discrete, finite range of values; the constraints are unary or binary predicates on these parameters; and there are no optimization criteria. In such a case, the constraint satisfaction methods of [9] apply. Similarly, other types of design tasks are well-fitted to other standard methods (integer programming, multi-objective optimization techniques, AND/OR graph search [26], numerical analysis techniques, etc.). Many of these conventional methods have performance guarantees of various sorts: linear programming and AND/OR graph search are guaranteed to find a global optimum; if the constraint network is a tree, constraint satisfaction methods are guaranteed to run in polynomial time; etc.

1.4.2.2. Knowledge-based routine design

Viewed as a knowledge-based search, a routine design process is comprised of several different types of basic operations: refinement, constraint processing, patching and optimization. *Refinement* and *implementation* operations generate new, and less abstract points in the search space; *constraint processing* operations prune inconsistent alternatives from consideration by the search; *patching* operations convert incorrect or sub-optimal designs into correct or more nearly optimal designs; *optimization* operations convert sub-optimal designs into designs that are more nearly optimal, with respect to some optimization criterion. Such operations might be stored as rules whose application requires pattern-matching (e.g., as in the VEXED system -- Volume I, Chapter 8); or as plans or procedures that are directly indexed by the type of design part to which they apply (e.g., as in the AIR-CYL system -- Volume I, Chapter 7).

1.4.2.3. Non-iterative, knowledge-based routine design

For some design tasks, sufficient knowledge or methods are available that a single pass (more or less) of top-down refinement -- possibly aided by constraint processing, patching, and directly applicable control knowledge -- is generally sufficient for converging on an acceptable design. This kind of design process model is demonstrated in several systems discussed in this book, including AIR-CYL (Volume I, Chapter 7) and VEXED (Volume I, Chapter 8). In the best case, applying this model requires running time linear in $p*l$, where p is the number of parts in the original specification, and l is the number of levels of abstraction through which each such part must be refined. However, constraint processing can slow things down, particularly if relatively global constraints are being processed [13].

1.4.2.4. Iterative, knowledge-based routine design

In other cases, the same kind of basic operations (refinement, constraint processing, etc.) are involved, but several (but not an exponential number of) iterations are generally required before an acceptable design is found. The need for iteration often arises when *multiple* constraints and objectives must be satisfied. A move in the design space that is good with respect to one constraint or objective may impair the satisfaction of another; tradeoffs may be necessary, and quickly finding a reasonable tradeoff (e.g., something close to a pareto-optimal solution) generally requires extensive domain-specific knowledge.

Several forms of iteration are possible:

- *Chronological backtracking.* A knowledge-poor method that is generally not acceptable for guaranteeing rapid convergence unless the density of solutions in the design space is very high, or the design space is very small. (Note, though, that "very small" need not mean a space of tens of designs, but -- given the speed of modern-day computing -- could be one containing thousands of designs. See, e.g., Volume I, Chapter 4, where an acceptable design for an automobile transmission is found using chronological backtracking.

- *Knowledge-directed backtracking.* Dependency-directed backtracking possibly aided by advice or heuristics. PRIDE (Volume I, Chapter 9) and VT (Volume I, Chapter 11) both illustrate this kind of iteration.

- *Knowledge-directed hillclimbing.* Iterative optimization or patching of a design until all constraint violations have been repaired, and an acceptable tradeoff has been met among all global optimality criteria (e.g., area, power consumption, delay time, in VLSI design). The knowledge used to select among different possible modifications could be an evaluation function, or a set of domain-specific heuristics (CHIPPE, Volume I, Chapter 12), or the choice could be made by the user (DESIGNER, Volume I, Chapter 14).

- *Knowledge-directed problem re-structuring.* It is not only possible to change the design solution but also the design problem, e.g., by adding new constraints or objectives, or retracting or relaxing old ones. As the original problem poser, the user is often made responsible for such changes [BIOSEP (Volume I, Chapter 10) and WRIGHT (Volume I, Chapter 13)].

In the best case, applying this model requires running time *polynomial* in $p*l$, where p is the number of parts in the original specification, and l is the number of levels of abstraction through which each such part must be refined; i.e., the number of iterations is polynomial in $p*l$. In the worst case, the number of iterations is exponential because whatever knowledge is guiding the search turns out to be inadequate or inappropriate.

1.4.2.5. Routine design systems covered in this volume

Table 1-4 classifies along the dimensions we have been discussing the various routine design systems described in later chapters of this book. Notice that most of these routine design systems address design tasks involving parameter value assignment or structure configuration (but not "from scratch" synthesis of the entire structure).

Table 1-4: Categorization of Systems and Methods
for Performing Routine Design

SYSTEM OR METHOD	DESIGN TASK	CHAPTER (VOL. I) OR PAPER	UNSPECIFIED STRUCTURE	DIRECTLY APPLICABLE KNOWLEDGE	SUBPROBLEM INTERACTIONS	ABSTRACTION LEVEL GAP
conventional optimization techniques	many simple tasks	--	parameter values	generative; control	algebraic constraints (global)	0
CSP methods	many simple tasks	[8]	parameter values	generative; some control	works best for semi-local constraints	0
AIR-CYL	air cylinders	7	parameter values	generative; patching	weak interactions	1
VT	elevators	11	parameter values	generative; knowledge-directed backtracking	strong interactions	0
PRIDE	copier paper paths	9	structure configuration	generative; knowledge-directed backtracking	works best for weak interactions	n
VEXED	circuits	8	entire structure	generative	weak interactions	n
BIOSEP	protein purification processes	10	structure configuration	generative	weak interactions + cost function	n
CHIPPE	VLSI	12	structure configuration	generative; knowledge-directed hillclimbing	weak interactions + global resource budgets	n
WRIGHT	spatial layouts	13	structure configuration	generative; user control	algebraic constraints + evaluation function	1
DESIGNER	graphic interfaces	14	structure configuration	generative; user control	mostly semi-local constraints	1

1.4.3. Models of Innovative Design

In innovative design tasks, routine design is not possible because of *missing design knowledge*. The missing knowledge might either be knowledge for directly generating new points in the design space, or knowledge for directly controlling the design space search. In this section, we will examine four different classes of innovative design. The first three focus (primarily) on missing generative knowledge, while the last deals with missing control knowledge:

- Innovation via case-based reasoning

- Innovation via structural mutation

- Innovation by combining multiple knowledge sources

- Search convergence by explicit planning of the design process

The first three approaches can be used to create innovative *designs*; the last approach involves creating innovative *design plans*, or innovative reformulations of the *design problem*.

1.4.3.1. Missing design knowledge

Why might relevant design knowledge be missing? One reason is that the most naturally acquirable knowledge might not necessarily be in a directly applicable form. This is often so in *case-based reasoning*; old designs and design process traces can be stored away fairly easily (if stored verbatim) in a case database, but then this leaves the problem of how to use these old cases to help solve a new design problem.

A second reason is that it generally is impossible to store the large amount of specific knowledge that would be necessary to adequately deal with all possible design variations (e.g., varying functional specifications, objective criteria, etc.). While some of this knowledge could be generalized, generalization often incurs a price of some sort; e.g., the generalized knowledge is not quite operational and must be made so at run-time; the (overly) generalized knowledge is not quite correct in all the circumstances to which it appears to be applicable; etc. Additionally, some of the knowledge simply is idiosyncratic, and thus not generalizable.

For this reason, deliberate engineering tradeoffs usually must be made in how much directly applicable design knowledge to build into the system, and how much to leave out, letting the system (or the user) cope with the missing knowledge.

A third reason is that human beings themselves may not have the relevant knowledge. Sometimes this is because the "structure to function" mapping is too complex to invert; methods may be available for analyzing the behavior and function of a given device, but not for taking a specified function and directly producing a structure that realizes that function. A case-based approach is often taken for such design tasks.

1.4.3.2. Case-based reasoning

Any case-based model of design must address the following issues:

- design case representation and organization
- design case storage
- design case retrieval
- design case adaptation and reuse

We will now say how three systems described in Volume II -- the BOGART circuit design system (Chapter 2), the ARGO circuit design system (Chapter 3), and the CADET system for designing fluid-mechanical devices (Chapter 4) -- handle these different issues. Chapter 5 (Volume II) analyzes case-based models of design in greater detail.

Design case representation. In BOGART, the stored cases are *design plans*, i.e., the steps used to incrementally refine a functional specification of a circuit into a pass transistor network are recorded *verbatim*. In ARGO, the same design session can yield several design cases, each at a different level of generality. Cases are stored as rules ("macrorules"), wherein the precise conditions for reuse of that case are stated explicitly. In CADET, each case involves four different representations: linguistic descriptions (i.e., <object attribute value> tuples); functional block diagramming; causal graphs; and configuration spaces.

Design case storage. In BOGART, the cases were automatically stored verbatim (when the user so chose) after a session with the VEXED design system (Volume I, Chapter 8). In ARGO, the design plan (a network of design steps and dependencies among them) is partitioned into levels. By dropping away more levels, more general plans are produced. Explanation-based generalization [19] of these design plans is used to determine the conditions under which each of these plans is applicable (which are then cached, along with the corresponding plans). In CADET, the cases were manually entered (since the focus of the CADET research was on case retrieval, and not case storage).

Design case retrieval. Because ARGO stores cases in such a way that the conditions for precise re-use are associated with them, retrieval of *applicable* cases is not an issue; ARGO uses a heuristic to restrict its retrieval to maximally specific cases. In BOGART, the *user* selects a case conceived as being similar to the current problem. In CADET, if no case directly matches the current specification, transformations are applied to the specification of device behavior until it resembles some case in the case database (e.g., some previously design artifact actually produces the desired behavior or something similar to it). In CADET, the specification may also be transformed in such a way that different parts of it correspond to different cases in the case database; all these cases are then retrieved (and the designs are composed).

Design case adaptation and reuse. In ARGO, reuse is trivial; a macrorule that matches is guaranteed to be directly applicable to the matching context. The transformations performed by CADET prior to retrieving a design permit direct use of the designs in the retrieved cases. In a case retrieved by BOGART (a design plan), some steps may apply to the current problem, while other parts may not; *replay* of the design plan is used to determine which steps apply. [23] is worth reading as a framework for case-based models of design such as BOGART, whose *modus operandi* is design plan replay.

Summary. BOGART's main innovation is in its method for design case reuse (via replay); ARGO's is in design case storage (macrorules with conditions of applicability); CADET's contribution is its method for design case retrieval (via transforming the design problem). All of these systems make contributions to the representation and organization of design cases that support their primary contribution.

1.4.3.3. Innovation via structural mutation and analysis

Most directly applicable knowledge for generating new points in the design space (either via refinement or modification) guarantees that something is being held invariant; most commonly, the functionality of the old design is preserved. If functionality-preserving transformations are not available, a weaker approach is to apply transformations that modify the artifact's (physical) structure in some manner, and then analyze the resulting functionality. Such analysis may then suggest further directions for modification until the desired functionality is (re)achieved. Such modifications are also guided by performance criteria and resource limitations.

One such approach is described in Volume II, Chapter 6. Here the problem is to find a way to simplify a given, modular design (modular in that each structural part implements a different function) by identifying and exploiting structure-sharing opportunities (i.e., ways to make a given structure achieve multiple functions). Here the transformation for modifying the artifact's structure is one that deletes some part of the structure. After a part has been deleted (and hence a function has been unimplemented), other features of the remaining structure are identified that can be perturbed to achieve the currently unimplemented function (while not ceasing to achieve the function(s) they are already implementing). The identified features are then perturbed in the direction of better achieving the unimplemented function. For example, the handle of a mug could be safely deleted if the remaining cylinder were sized and shaped in such a way that it could be grasped by a human hand easily, and were made of a material that was heat-insulating (and hence would not burn the hand) -- e.g., a styrofoam cup. Essential to this approach is knowledge that associates changes in particular physical features of an artifact to the functions these (might) achieve.

If associations between (change of) physical structure and (change of) function are not hardcoded, then they may have to be derived. Qualitative modeling and reasoning of various kinds (e.g., qualitative simulation: see Volume II, Chapter 10) can sometimes be used to derive such associations.

1.4.3.4. Exploiting multiple knowledge sources

We have just described systems that use a case database to generate new designs, and other systems that use associations between structure and function to do the same. For some design tasks, multiple sources of (such indirectly usable) knowledge may be available, and all potentially useful; it might even be the case that solving the design problem *requires* integrating the advice of several knowledge sources.

Chapter 7 (Volume II) describes the ALADIN system, which helps design new aluminum alloys that meet specified properties. ALADIN draws on several sources of expertise to generate new points in the design space:

- a case database of previously designed alloys and their properties
- if-then rules which associate structural changes (e.g., adding magnesium to the alloy) with functional changes (e.g., the value of the "strength" property increases).
- mathematical models of physical properties
- statistical methods for interpolation and extrapolation

1.4.3.5. Planning the design process

In a simple *routine design* scenario, the control questions that must be answered along the way take relatively simple forms: which part of the design to work on next? What to do there (refine, implement, optimize, patch)? Of several possible ways to do that, which to pick? Acquirable control knowledge may be sufficient for answering the control questions as they arise.

However, for several reasons, a design process model can be more complex, thus giving rise to new control questions, and hence to the need for a more complex controller:

- *More methods and knowledge sources.* Innovative design systems can involve a diverse range of activities and draw on many sources of knowledge. For example, the ALADIN system draws on multiple knowledge sources, and consequently must also answer new control questions: which knowledge source to consult next? How to combine the outputs of several knowledge sources? etc.

- *Multiple objectives.* Another source of control problems arises when multiple objectives must be satisfied. New control questions include: With respect to which objective should the design be improved next? Which part of the design should be redesigned to effect the improvement?

- *Expensive design operations.* Operations such as simulation (e.g., VLSI chip simulation) or analysis (e.g., finite element analysis) can be sufficiently costly that their use should be carefully planned.

A global view: Control as planning. To be operational, any control strategy must provide answers to specific, *local* control questions of the kind just described. However, the problem of control has a *global* goal in mind: Utilize knowledge and methods so as to most rapidly converge on an acceptable solution. Hence we can think of the problem of control as a *planning problem*: construct a relatively short design plan whose steps invoke various design methods and draw on design knowledge, and which, when completely executed, results in the creation of an acceptable design.

Stefik [37, 36] and Wilensky [45] gave the name *meta-planning* to this approach to control, since the design process itself is being explicitly represented and reasoned about. Stefik's MOLGEN system represented the design (a plan for a molecular genetics experiment) at multiple levels of abstraction. MOLGEN took a least commitment approach to refining the design through these levels of abstraction. It also used a multi-layered control strategy, explicitly representing and modifying the design plan. The ALADIN system (Volume II, Chapter 7) uses a very similar approach to managing the navigation through its multiple spaces for designing aluminum alloys.

Control as top-down refinement of design plans. When design operations (such as VLSI simulation) are expensive, one response is to create abstractions of these operations and much more cheaply construct plans for the design process in the space of abstract operations, pick the best abstract plan, and then refine it into an actual design plan (one whose execution would produce complete designs, and accurate analyses). This approach can be viewed as a special kind of meta-planning in which the planning method is top-down refinement (often also called "hierarchical planning"). This approach has been applied to VLSI design in the ADAM system (Volume II, Chapter 8).

But what is the "best" abstract plan? In ADAM, "best" means the one which when executed, creates a design that comes closest to satisfying all of several resource limitations (on area, speed, power, and design time). ADAM uses a single weighted evaluation function of all the resource usages:

```
w1 * area + w2 * speed + w3 * power + w4 * design time

where w1+w2+w3+w4=1
```

to guide its search. ADAM first finds plans that construct designs which are optimal with respect to each of the individual resources; for instance, to do so for "area" would involve setting $w1 = 1$, and $w2 = w3 = w4 = 0$. Based on the the difference between the costs of the resulting designs and the specified budgets, ADAM uses *linear interpolation* to readjust the weights on the evaluation function. It then replans.

Exploratory design: Control as hillclimbing in the space of problem formulations. The following hypothesis (we will call it the *routine design hypothesis*) is one way of viewing the relationship between an innovative design problem and a routine design problem:

> If the design problem is appropriately structured and contains enough detail (i.e., if we are "looking at the problem right"), then a single pass of a simple routine design process should produce an acceptable design (if one exists).

The control strategy we will next describe, called *exploratory design*, is appropriate for those problems where the initial design problem is *not* appropriately structured or annotated (i.e., it is an *innovative* design problem). We call this "exploratory design" because our intuition is that human designers handle problems that are complex in novel ways by spending their initial time finding a good way to look at the problem.

Models of routine design involve a search purely in the space of designs. In exploratory design, the problem and the solution co-evolve. Exploratory design hillclimbs in the space of problem formulations (the "outer loop" of the method), getting feedback for adjusting the problem formulation from analyzing how the candidate designs generated so far (by the "inner loop" of routine design) fail to be acceptable.

The DONTE system (Volume II, Chapter 9) performs such hillclimbing in the space of circuit design problem formulations using top-down refinement, constraint processing, and design patching operations in its "inner loop". The kind of problem reformulation operations it performs there are: macro-decision formation, which imposes a hierarchical structure on a relatively flat problem decomposition; budgeting, which adds a new budget constraint to every design component; re-budgeting, which may adjust such constraints in several components; rough design, which assigns estimates of resource usage to various parts of the design; and criticality analysis which (re)assesses how (relatively) difficult the various subproblems are to solve (given their current budgets, etc.).

1.4.3.6. Innovative design systems covered in this volume

Table 1-5 classifies along the dimensions we discussed earlier the various innovative design systems described in later chapters of this book. Notice that most of these innovative design systems address design tasks involving synthesis of the entire structure.

1.4.4. Qualitative Reasoning about Artifacts during Design

The mapping of a knowledge level specification of a design system into am algorithm level search algorithm can draw on formally represented bodies of generally useful "common sense" knowledge and procedures relevant to reasoning about the physical artifacts being designed. We now describe two kinds of such knowledge: knowledge about physical systems; and knowledge about geometry. With respect to codification of "common sense" knowledge, the CYC project [14] represents an alternate and possibly complementary approach to those described here.

Table 1-5: Categorization of Systems and Methods
for Performing Innovative Design

SYSTEM OR METHOD	DESIGN TASK	CHAPTER (VOL. II)	UNSPEC. STRUC.	ABSTR. LEVEL GAP	GENERATION PROBLEMS ADDRESSED	CONTROL PROBLEMS ADDRESSED	WHAT IS INNOVATIVE
BOGART	circuits	2	entire structure	1	how to replay retrieved case		design
ARGO	circuits	3	entire structure	1	how to store cases so generation is easy		design
CADET	fluid-mechanical devices	4	entire structure	n	how to identify similar cases		design
FUNCTION SHARING	fluid-mechanical devices	6	none	0	how to identify function-sharing possibilities		design
ALADIN	aluminum alloys	7	entire structure	n spaces	how to use multiple knowledge sources to generate new design		design
ADAM	VLSI	8	entire structure	n		how to find promising design plan	design plan
DONTE	circuits	9	entire structure	n		how to find good problem decomposition, budget allocation, resource usage estimations	design problem reformulation

1.4.4.1. Qualitative reasoning about physical systems during design

Functional specifications for *physical systems* often take the form of stipulating a particular relationship between behavioral parameters, e.g., the output rotation of a rotation transmitter must be 30 times as fast as the input rotation. It is rarely the case that a single part (e.g., a single gear pair) is capable of directly achieving the specified relationship. Instead, a series of interacting components may be needed. This is especially the case when the type of the behavioral parameter changes: e.g., the input is a rotational speed, but the output is a rate of up-and-down movement. The network of interacting behavioral parameters may necessarily include feedback loops, e.g., when the specified relationship defines a self-regulating device (e.g., a change in one variable should result in a corresponding change in the other).

Williams has proposed a design process model for such problems called *interaction-based invention*:

> Invention involves constructing a topology of interactions that both produces the desired behavior and makes evident a topology of physical devices that implements those interactions [46].

One of the key steps in this process is verifying that the interactions in the constructed interaction topology actually "compose" to produce the specified interaction. Carrying out this step (and satisfying its representational needs, i.e., providing an adequate representation of the causal and temporal features of each interaction) is particularly difficult when the topology is complex (e.g., as in most circuits that contain feedback loops). Chapter 10 (Volume II) discusses how to adequately represent such interactions in complex physical systems (such as analog circuits with feedback loops), and how to qualitatively analyze the global behavior of these systems.

1.4.4.2. Qualitative reasoning about geometry in design

Geometry-constrained synthesis. Many design tasks involve geometry in one way or another in their functional specifications or domain knowledge. In the simplest of cases, the role geometry plays is purely static, placing restrictions on the boundaries of the artifact, points of attachment of parts of the artifact, etc. The WRIGHT system described in Chapter 13 (Volume II) handles a subclass of such spatial placement problems.

The synthesis of small load-bearing structures illustrates a more complex role of geometry: *forces* (i.e., the loads) are positioned at certain points in space; a single structure must be synthesized that is both stable and capable of bearing the loads (and that does not occupy any "obstacle" regions of space). Chapter 11 (Volume II) describes the MOSAIC system, which synthesizes such load-bearing structures using a design process model that performs problem abstraction, problem decomposition, and iterative re-design.

Another geometric complication shows up in *kinematic synthesis*, the synthesis of physical structures that *move* in ways that satisfies certain restrictions on motion in space. Chapter 12 (Volume II) considers the problem of designing linkages (e.g., door hinges, aircraft landing gear, cranes, etc.), given constraints on specific points through which the linkage must pass (perhaps in a particular order), number of straight line segments in the path of motion, etc. In the TLA system, the user selects a linkage from a case database of four-bar linkages, looking for those that have features resembling the problem specifications. Optimization techniques are then used to adapt the known case to the current problem; user intervention helps such techniques avoid getting stuck in local minima.

Joskowicz (Volume II, Chapter 13) also describes an approach to kinematic synthesis. Mechanisms, retrieved from either a catalog or a case database, are considered during artifact redesign. Retrieved mechanisms should ideally be *kinematically equivalent* to the current design. Joskowicz describes a method for comparing two mechanisms for kinematic equivalence, that involves trying to find a common abstraction of both. This same mechanism comparison technique is used to organize the case database (for the purpose of efficient retrieval) into classes of kinematically equivalent mechanisms.

Geometry-based analysis. That designed artifacts have geometric features means that some of the analysis processes performed during design will involve geometric reasoning, including: static and dynamic analysis of stresses (based on shape), and kinematic simulation of mechanisms.

The conventional approach to analyzing stress is finite element analysis. However, this method requires a grid as an input, and which grid is best varies with the problem. In contrast, Chapter 14 (Volume II) describes an approach to stress analysis that geometrically partitions an object into regions in such a way that the object parts have shapes (e.g., a plate with a hole in it) resembling known cases (e.g., a plate without a hole in it). These known cases have associated (pre-computed) stress analyses, which are then used as part of the stress analysis data for the overall object.

One method for *kinematic simulation* is described in Chapter 13 (Volume II). First, local behaviors are computed from two-dimensional configuration spaces, defined by the objects' degrees of freedom. Global behaviors are then determined by composing pairwise local behaviors.

1.5. BUILDING A KNOWLEDGE-BASED DESIGN TOOL

The actual construction of a new knowledge-based design tool goes through three basic phases:

- Identify the design task
- Configure and instantiate the design process model

• Implement the design process model

1.5.1. Identifying the Design Task

Identifying the design task involves defining the task and classifying it.

1.5.1.1. Knowledge acquisition to define the design task

To define a design task, we must acquire knowledge defining:

• the class of *problems* that can be solved;

• the class of *candidate solutions* that contains a set of acceptable solutions to the problem;

• the *domain theory*, the body of domain-specific knowledge that is accessed in solving such problems, and constrains what is considered to be an acceptable solution.

How can such design knowledge be either easily acquired from domain experts, or otherwise automatically added to the knowledge base?

Graphical interfaces. Chapter 2 (Volume III) discusses the advantages of using graphical interfaces in acquiring design knowledge from experts. In particular, the knowledge is acquired in the form of decision trees. These trees are then mapped into expert rules in OPS5. The complete process is illustrated by acquiring and compiling knowledge from experts for bearing selection.

Knowledge acquisition for specific design process models. Another way to simplify knowledge acquisition is to tailor a particular knowledge acquisition method to a specific design model. For example, the SALT system (Volume I, Chapter 11) specializes in acquiring knowledge for a design system that iteratively modifies a design.

SALT first acquires a graph whose nodes are design inputs, design parameters, or design constraints and whose edges express various relationships between these. SALT then acquires three types of knowledge that are indexed off the graph nodes: knowledge for proposing a design extension (specifying a design parameter), knowledge for identifying a constraint, and knowledge for proposing a fix to a constraint violation. SALT has a schema for each type of knowledge, and prompts the user with questions whose answers fill in the appropriate schema. SALT also has techniques for analyzing the completeness and consistency of the knowledge base. The SALT system was used to acquire the knowledge in the VT system.

Case-based reasoning. In Section 1.4.3.2, we described case-based reasoning as a particular model of innovative design. Because case-based reasoning involves storage of design cases from previous design system sessions, it represents another way of adding "new" knowledge to the knowledge base.

As mentioned previously, the stored knowledge can range in generality from design plans that are stored verbatim (as in the BOGART system, Volume II, Chapter 2), to automatically generalized knowledge (as in the ARGO system of Volume II, Chapter 3).

1.5.1.2. Classifying a design task

As mentioned earlier, design tasks can be classified along several dimensions, including:

- available methods and knowledge
- gap in abstraction levels between specification and implementation
- amount of unspecified (physical) structure
- complexity of interactions between subproblems; and
- amount and type of knowledge a system user can provide

1.5.2. Configuring and Instantiating the Design Process Model

Classification of a design task identifies important features of that task. Different features suggest different design process models. Tables 1-4 and 1-5 suggest, by example, some of the correspondences.

1.5.3. Implementing the Design Process Model

Once a design process model is determined, the next step is to map the design process model onto the program level (see Figure 1-1). "Maxims" pertinent to carrying out this mapping include:

1. Code in an appropriate programming language, such as C++, LISP, OPS5, KEE^{TM}. Most of the papers in Volume I and Volume II, as well as Chapter 7 in Volume III, take this approach.

2. Use a commercial tool that provides some support for design artifact representation; implement appropriate extensions. Chapters 3, 4, 5, and 6 in Volume III follow this path.

3. Develop a domain-independent shell that implements the design process model(s) and instantiate the shell for a particular application.

4. Use a knowledge compiler to generate special-purpose procedures for efficiently processing particular (and generally domain-specific) subtasks of the overall design task.

1.5.3.1. Commercially available tools

There are two kinds of tools available in the commercial market place for civil/mechanical engineering applications (see Table 1-2):

1. **Parametric modelers,** which provide constraint processing capabilities to geometric modelers. An application utilizing a parametric modeler (DesignView^{TM}) and a knowledge-based programming tool (NEXPERT^{TM}) for designing a product and forming sequence for cold forging is described in Chapter 4 (Volume III). We have included a list of commercial tool vendors in Appendix A at the end of this chapter.

2. **Design representation frameworks,** which provide additional layers over knowledge representation languages. Typically these layers support the following activities:

   • Representation of engineering entities, including composite objects;
   • Geometric modeling;

- Constraint management;
- Access to external programs, such as engineering databases;
- Programming language support (current tools are implemented in LISP); and
- Rule-based inferencing.

Applications implemented in three commercially available tools are described in Volume III, Chapters 3 (ICAD^{TM}), 4 (DesignView^{TM} and NEXPERT Object^{TM}), 5 (Design++^{TM}), and 6 (Concept Modeller^{TM}).

1.5.3.2. Domain-independent shells

Domain-independent shells, in addition to representation and programming language support, provide design process models as problem solving strategies. Applications can be built by adding domain-specific knowledge. Many of the routine design systems described in Volume I have evolved into domain-independent shells. These systems view design as:

```
Hierarchical Refinement + Constraint Propagation + ..
```

and provide knowledge editing facilities for inputting design plans, goals, artifacts, and constraints. Table 1-6 summarizes several domain-independent shells.

1.5.3.3. Knowledge compilers

In principle, *knowledge compilers* can be used to create (at compile time) those components of the design system that are not easily viewable as instantiations of domain-independent "shell" components, and that are not one of the commercially available tools (e.g., parametric modellers or design representation frameworks). Often the compiled components handle particular, domain-specific tasks such as maze routing [32], house floorplanning [44], or synthesis of gear chains [24]. It is also possible to use knowledge compilers to optimize components that originated as shell instantiations.

Some compilers are quite specialized; for example, the ELF system [32] specializes in compiling global routers, for varying VLSI technologies. The KBSDE compiler [44] and the constraint compiler of the WRIGHT system

Table 1-6: Domain-Independent Shells that Implement
Hierarchical Refinement and Constraint Propagation

SHELL/ REFERENCE	PREDECESSOR/ DOMAIN	REP. LANGUAGE/ BASE LANG.	MACHINE OR OS	DEPARTMENT/ PLACE
DESCRIBE [20]	PRIDE Paper Handling	LOOPS LISP	XEROX	Only Inhouse
EDESYN [16]	HI-RISE Buildings	FRAMEKIT LISP	Unix	Civil Engrg. CMU
DSPL [4]	AIR-CYL Air Cylinders	LISP	Unix	Comp. Sci. OSU & WPI
EVEXED [38]	VEXED VLSI	STROBE LISP	XEROX	Comp. Sci. Rutgers
DIDS [2]	MICON Computers	C++ C	Unix	EECS Univ. Michigan
CONGEN [34]	ALL-RISE Buildings	C++ C	Unix	Civil Engrg. M.I.T.

(Volume I, Chapter 13) address a different and somewhat broader class of knowledge-based systems for spatial configuration tasks. The DIOGENES compiler [24] addresses the still broader class of heuristic search algorithms. These compilers appear to obey the standard power/generality tradeoff. The models of knowledge compilation also grow progressively weaker as the breadth widens, culminating in such weak (i.e., relatively unrestricted) models as: a transformational model of knowledge compilation [22] or a model of knowledge compilation as formal derivation.

All the compilers just mentioned are research prototypes, and are thus not commercially available. Nonetheless, we mention this technology because of its potential importance in the not too distant future. In the meantime, human programming skills will have to suffice.

1.6. DESIGN AS PART OF A LARGER ENGINEERING PROCESS

It is important to view design in the perspective of the overall engineering process, which involves several phases: market studies, conceptualization, research and development, design, manufacturing, testing, maintenance, and marketing (see Figure 1-3). In this process people from various disciplines interact to produce the product.

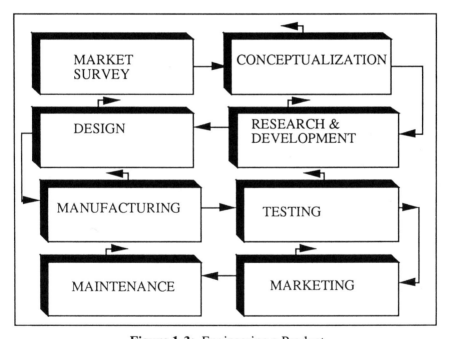

Figure 1-3: Engineering a Product
(Bent arrows indicate that the process is iterative)

In traditional product development, the lack of proper collaboration and integration between various engineering disciplines poses several problems, as expounded by the following Business Week (April 30, 1990, Page 111) clip [see Figure 1-4 for a typical scenario in the AEC industry].

The present method of product development is like a relay race. The research or marketing department comes up with a product idea and hands it off to design. Design engineers craft a blueprint and a hand-built prototype. Then, they throw the design "over the wall" to manufacturing, where production engineers struggle to bring the blueprint to life. Often this proves so daunting that the blueprint has to be kicked back for revision, and the relay must be run again - and this can happen over and over. Once everything seems set, the purchasing department calls for bids on the necessary materials, parts, and factory equipment -- stuff that can take months or even years to get. Worst of all, a design glitch may turn up after all these wheels are in motion. Then, everything grinds to a halt until yet another so-called engineering change order is made.

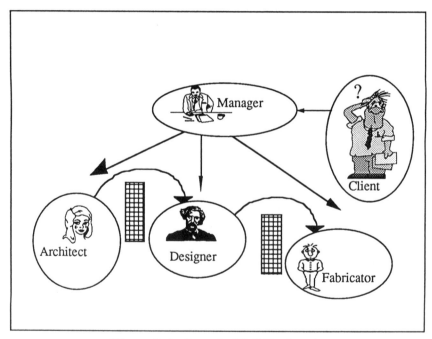

Figure 1-4: Over the Wall Engineering

Several companies have addressed the above problem by resorting to a more flexible methodology, which involves a collaborative effort during the entire life cycle of the product. It is claimed (Business Week, April 1990) that this approach[2] results in reduced development times, fewer engineering changes, and better overall quality. The importance of this approach has been recognized by the Department of Defense, which initiated a major effort -- the DARPA Initiative in Concurrent Engineering (DARPA DICE) -- with funding in the millions of dollars.

It is conceivable that the current cost trends in computer hardware will make it possible for every engineer to have access to a high performance engineering workstation in the near future. The "over the wall" approach will probably be replaced by a network of computers and users, as shown in Figure 1-5; in the figure we use the term *agent* to denote the combination of a human user and a computer.

The following is a list of issues that we consider important for computer-aided integrated and cooperative product development.

1. **Frameworks**, which deal with problem solving architectures.

2. **Organizational issues**, which investigate strategies for organizing engineering activities for effective utilization of computer-aided tools.

3. **Negotiation techniques**, which deal with conflict detection and resolution between various agents.

4. **Transaction management issues**, which deal with the interaction issues between the agents and the central communication medium.

5. **Design methods**, which deal with techniques utilized by individual agents.

6. **Visualization techniques**, which include user interfaces and physical modeling techniques.

Several papers in Volume III address some of the above issues; [33] contains additional papers in this area. Chapters 7 and 8, Volume III, discuss the DFMA and the ECMG frameworks, respectively, that bring manufacturability knowledge into the early design phases. The manufacturing knowledge is tightly integrated into the the design framework. The Engineous system, described in Volume III, Chapter 9, is a generic shell that combines knowledge-

---

[2]"Concurrent engineering", "collaborative product development", "cooperative product development", "integrated product development" and "simultaneous engineering" are different phrases used to connote this approach.

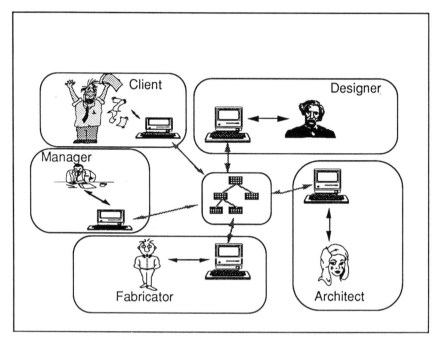

Figure 1-5: Modern View of Product Development

based expert systems, numerical optimization, and genetic algorithms for product design.

While the above systems are closely coupled architectures, the systems described in Chapters 10, 11, and 12 (Volume III) are loosely coupled and reflect the architecture shown in Figure 1-5. A multi-level and a multi-modal architecture, DMA, that supports easy integration of various design/manufacture CAD systems is proposed in Chapter 10 (Volume III). The design module supports an axiomatic approach to design [41]. The manufacture module contains manufactability knowledge, such as assembly sequencing, etc.

A dual design partner scheme is proposed in Chapter 11 (Volume III). This scheme supports two competing system behaviors. One expert machine -- the stabilizer -- resists change and always presents a conservative hypothetical model of the product. The other expert machine -- the innovator -- strives for well calculated and justified alternative hypothetical models of the product. The dual partner scheme is being implemented using the blackboard architecture [25].

The DICE project (Volume III, Chapter 12) implements a blackboard architecture over an object-oriented database management system; thus the blackboard and the object-store are tightly integrated. In addition, the objects in the blackboard have behavior associated with them. Hence, the need for a sophisticated scheduler -- as provided in the traditional blackboard systems -- is obviated. The DICE project also incorporates comprehensive transaction and version management mechanisms. The DICE version described in this volume was implemented in Common LISP. Other implementations also exist in the OPAL/GEMSTONE and C++/ONTOS environments. Table 1-7 summarizes the various efforts in integrated design systems.

Table 1-7: Summary of Integrated Design Frameworks

SYSTEM	CHAPTER (VOL. III)	FEATURES	NO. LEVELS	STATUS
DFMA	7	Tightly coupled	1	In-house use
ECMG	8	Tightly coupled; Domain-independent	1	Commercially available
Engineous	9	Tightly coupled expert systems; genetic algorithms; optimization	1	In-house use
Dual Partner	11	Loosely coupled; Blackboard; database	n	Prototype
DMA	10	Loosely coupled	n	Prototype
DICE	12	Loosely coupled; Blackboard; object-oriented databases; negotiation; transaction management	n	Prototype

1.7. SUMMARY

In this overview chapter, we have presented a framework for helping to understand the field of "AI in Engineering Design" in general, and the papers in this collection, in particular.

Applying AI software engineering methodology to Engineering Design problems. We first considered "Engineering Design" and "Artificial Intelligence" as separate disciplines, the former providing special kinds of ill-structured problems, and the latter providing a methodology for developing knowledge-based systems that effectively solve certain types of ill-structured problems.

Design problems are *ill-structured* in that the mapping of desired functionality onto a (physical) structure that correctly implements it is generally not straightforward. Furthermore, most design problems call for not only a *correct* design but a *good* design -- good with respect to one or more (possibly ill-defined) metrics (e.g., cost, area, volume, power consumption, etc.); this further complicates the mapping, thereby decreasing the likelihood that a simple (polynomial time) algorithm will suffice for carrying out the mapping, and increasing the likelihood that some degree of search (e.g., generate-and-test) will be necessary. Finally, the design problem representation itself may begin its life as an ill-structured set of "requirements" and only gradually (enabled by feedback from actual design experience) evolve into a set of formal "specifications".

For the purposes of this book, we have described Artificial Intelligence as a discipline that provides a multi-level methodology for engineering knowledge-based problem-solving systems. In particular, a *knowledge level* specification of the system (and the class of problems it must solve) is mapped into an *algorithm level* description of an efficient search algorithm for efficiently and acceptably solving that class of problems. That (simulatable) algorithm description is then mapped into an actual piece of code at the *program level,* using one or more programming paradigms (e.g., procedural programming, rule-based programming, object-oriented programming), shells (e.g., VP-EXPERT^{TM}), or commercially available subsystems (e.g., an ATMS in KEE^{TM}). The application of AI to Engineering Design thus looks like a specialization of this software engineering methodology to: design tasks (specified at the "knowledge level"); design process models (described at the "algorithm level"); and design programs built from shells, commercially available design subsystems, and manually constructed code (implemented at the "program level").

Mapping a knowledge level specification for a design system into a algorithm-level search algorithm. In considering mapping a knowledge level specification for a design system into an algorithm-level search algorithm, it is useful to decompose the algorithm into passive and active components. One passive component is the *design space* to be searched. The active design components are the various *functional components of the design process model* (e.g., refinement, hillclimbing, constraint propagation, backtracking, etc.), which, in effect, generate the design space and navigate through it. These active components draw upon another passive component, declaratively represented *design knowledge*, interpreting this knowledge at run time (e.g., to estimate the cost of a particular design, to choose between several design alternatives, etc.).

The same piece of knowledge can be embedded into an algorithm in a variety of ways, with varying degrees of effectiveness. The most effective way to map available design knowledge into the algorithm-level search algorithm is to carefully engineer the design space itself, so that it -- *a priori* -- will contain (when generated at run-time) as few incorrect or poor designs as possible. The next most effective way to use design knowledge is to *compile* it into the active components of the search algorithm (e.g., creating customized routines for efficiently performing special tasks such as routing, placement, estimation, simulation, etc.) The least effective (though sometimes easiest, and sometimes necessary) way to use design knowledge is to represent it declaratively (e.g., as is often the case in shells), and then *interpret* it at run time.

Other factors also come into consideration when mapping a knowledge level specification of a design system into an algorithm-level search algorithm. Design tasks can be categorized along various dimensions; different search algorithms will be appropriate for different types of design tasks. Useful dimensions for taxonomizing design tasks include: available methods and knowledge (addressing that task); gap in abstraction levels between specification and implementation; amount of unspecified (physical) structure; and amount and type of knowledge a system user can provide.

Of primary importance in distinguishing types of design tasks is the amount and types of available knowledge (and the form in which the knowledge is available). The more design knowledge available in the right form, the more *routine* (or "direct") a design process can be used (involving a top-down refinement and/or hillclimbing process that converges on an acceptable design with little or no search). Any missing knowledge or knowledge in the wrong form or incorrect knowledge must be compensated for. Such *innovative* design problems can be addressed by various "indirect" techniques such as case-based reasoning, structural mutation, combining multiple knowledge sources, and explicit planning of the design process.

Design processes can be non-routine and *indirect* in the sense that generating new points in the design space may require an explicit problem-solving process, rather than the direct application of a single procedure or the direct interpretation of a single piece of knowledge. Using case-based reasoning to generate new points in the design space is usually indirect in that it requires nontrivial processes of design case selection, adaptation, and reuse. Using structural mutation to generate new points can be indirect in the sense that the quality and even the functionality of the mutations may not be knowable *a priori*, may require a problem-solving process (e.g., qualitative or numerical simulation) to determine, and may lead to a search through the space of possible mutations for a correct and good one. Using multiple knowledge sources to generate new points in the design space is usually indirect in that integrating partial solutions is a nontrivial problem-solving process.

Design processes can also be non-routine and indirect in the sense that *control* of the search is indirect -- it requires an explicit problem-solving process, rather than merely the direct application of a simple control procedure or the direct interpretation of a single piece of control knowledge to decide what to do next. The design search control problem can be usefully viewed as a *planning problem*, and various planning techniques can be applied: forward or backward planning, "hierarchical planning" (i.e., top-down refinement of design plans), or "exploratory design" (i.e., hillclimbing in the space of problem formulations).

The mapping of a knowledge level specification of a design system into an algorithm-level search algorithm can draw on formally represented bodies of generally useful "common sense" knowledge and procedures relevant to reasoning about the physical artifacts being designed. Much has been learned regarding qualitatively reasoning about *physical systems* in general. We have initial answers to such questions as: how to qualitatively simulate certain classes of physical systems; how to derive aggregate system behavior from the behavior of the parts; how to determine the function of the system given its aggregate behavior and a description of the system's context; etc. Much also has learned about (qualitatively) reasoning about the *geometry* of physical objects in general: how to satisfy placement and sizing constraints; how to satisfy constraints involving forces being applied at various points in space; how to satisfy kinematic constraints on how physical structures can move; how to analyze stresses based on shape; and how to simulate a mechanism's movement through space.

Mapping an algorithm-level search algorithm into a program. Implementing a design search algorithm can involve several types of tasks: coding in an appropriate programming language, such as C++, LISP, OPS5, KEE^{TM}; using commercially available tools for representing design artifact representations (e.g., parametric modellers) and for processing common tasks (e.g., constraint managers, geometric modellers and constraint managers, engineering databases); instantiating a domain-independent, design process shell (e.g., for hierarchical refinement and constraint propagation); and creating customized procedures or algorithms for special purpose tasks, either by hand, or by running a knowledge compiler.

Design as part of a larger engineering process. Design is only one phase or aspect of a larger engineering process that also includes market studies, conceptualization, research and development, manufacturing, testing, maintenance and marketing. The more the design process can be integrated with the other engineering phases, the more cost-effective the entire process will be. Approaches to computer-aided support of an integrated engineering process can range from loose couplings of the phases (facilitated by electronic mail, or shared files, or blackboard architectures), to tight couplings that constrain earlier phases (e.g., design) with requirements anticipated in later phases (e.g., manufacturing constraints) and reformulated so that they are expressed in the language of the earlier phases.

Other summary references. We have intended this chapter as a brief but complete summary of the state of the field of AI in Engineering Design. Other useful summary references worth reading include [3] (which introduced the "routine", "innovative", and "creative" design distinction), [21] (which distinguishes different design process models on the basis of types of design goal interactions), and [43] (which introduced the distinction between the "program level" and the "algorithm level", which was called the "function level" in that paper).

1.8. APPENDIX A: VENDORS OF AI-BASED TOOLS FOR COMPUTER-AIDED ENGINEERING

Ashlar, Inc.
1290 Oakmead Pkwy.
Sunnyvale, CA 94806
Tool: Vellum^{TM}

Integraph Corp.
Mail Stop WYLE3
Huntsville, AL 35894-0001
Tool: MicroStation^{TM}

Cognition Inc.
900 Tech Park Drive
Bellerica, MA 01821
Tool: ECMG^{TM} and MCAE^{TM}

Mentor Graphics
8500 South West Creek Side Place
Beaverton, OR 97005
Tool: ADE^{TM}, Logic Synthesizer^{TM}

ICAD Inc.
1000 Massachusetts Avenue
Cambridge, MA 02138
Tools: ICAD^{TM}

Parametric Technology Corp.
128 Technology Sr.
Waltham, MA 02154
Tool: Pro/ENGINEER^{TM}

ComputerVision
55 Wheeler Street
Cambridge, MA 02138
Tool: DesignView™

Spatial Technology
2425, 55th Street, Bldg. A
Boulder, CO 80301
Tool: ACIS™

Wisdom Systems
Corporate Circle
30100 Cagrin Blvd.
Suite 100
Pepper Pike, OHIO 44124
Tool: Concept Modeller™

1.9. REFERENCES

[1]   Benjamin, P., Ed., *Change of Representation and Inductive Bias*, Kluwer Publishing, 1990.

[2]   Birmingham, W. and Tommelin, I., "Towards a Domain-Independent Synthesis System," in *Knowledge Aided Design*, Green, M., Ed., Academic Press, 1991.

[3]   Brown, D. and Chandrasekaran, B., "Expert systems for a class of mechanical design activity," *Proc. IFIP WG5.2 Working Conf. on Knowledge Engineering in Computer Aided Design*, IFIP, September 1984.

[4]   Brown, D. and Chandrasekaran, B., *Design Problem Solving: Knowledge Structures and Control Strategies*, Morgan Kaufmann, San Mateo, CA, 1989.

[5]   Chandrasekaran, B., "Design Problem Solving: A Task Analysis," *AI Magazine*, 1990.

[6]   Clancey, W., "Classification problem-solving," *AAAI*, August 1984.

[7]   De Kleer, J., "An assumption-based TMS," *Artificial Intelligence*, Vol. 28, No. 2, pp. 127-162, March 1986.

[8]   Dechter, R. and Pearl, J., "Network-based heuristics for constraint satis-
        faction problems," *Artificial Intelligence,* Vol. 34, pp. 1-38, 1988.

[9]   Dechter, R., "Enhancement Schemes for Constraint Processing: Back-
        jumping, Learning, and Cutset Decomposition," *Artificial Intelligence,*
        Vol. 41, pp. 273-312, January 1990.

[10]   Feigenbaum, E. and Feldman, J., Ed., *Computers and Thought,* McGraw-
        Hill, New York, 1963.

[11]   Gero, J., Ed., *Preprints of the international round-table conference on
        modelling creativity and knowledge-based creative design,* University of
        Sydney, 1989.

[12]   Coyne, R., Rosenman, M., Radford, A., Balachandran, M., and Gero, J.,
        Knowledge-Based Design Systems, Addison-Wesley, Reading, Mass.,
        1990.

[13]   Kelly, Kevin M., Steinberg, Louis I., and Weinrich, Timothy M.,
        Constraint Propagation in Design: Reducing the Cost, unpublished
        working paper, March 1988, [Rutgers University Department of Com-
        puter Science AI/VLSI Project Working Paper No. 82].

[14]   Guha, R. and Lenat, D., "Cyc: A Mid-Term Report," *The AI Magazine,*
        Vol. 11, No. 3, pp. 32-59, Fall 1990.

[15]   Lowry, M. and McCartney, R., Ed., *Automated Software Design,* MIT
        Press, Cambridge, MA 02139, 1991.

[16]   Maher, M. L., "Engineering Design Synthesis: A Domain-Independent
        Approach," *Artificial Intelligence in Engineering, Manufacturing and
        Design,* Vol. 1, No. 3, pp. 207-213, 1988.

[17]   McCarthy, J. and Hayes, P., "Some philosophical problems from the
        standpoint of artificial intelligence," in *Readings in artificial
        intelligence,* Webber, B. and Nilsson, N., Ed., Morgan Kaufmann, Los
        Altos, CA., 1981.

[18]   Meyer, E., "Logic Synthesis Fine Tunes Abstract Design Descriptions,"
        Computer Design, pp. 84-97, June 1 1990.

[19]   Mitchell, T. M. and Keller, R. M. and Kedar-Cabelli, S. T.,
        "Explanation-Based Generalization: A Unifying View," *Machine
        Learning,* Vol. 1, No. 1, pp. 47-80, 1986.

[20]   Mittal, S. and Araya, A., "A Knowledge-based Framework for Design,"
        Proceedings AAAI86, Vol. 2, Philadelphia, PA, pp. 856-865, June 1986.

[21]   Mostow, J., "Toward better models of the design process," *AI
        Magazine,* Vol. 6, No. 1, pp. 44-57, Spring 1985.

[22]   Mostow, J., "A Preliminary Report on DIOGENES: Progress towards Semi-automatic Design of Specialized Heuristic Search Algorithms," *Proceedings of the AAAI88 Workshop on Automated Software Design,* St. Paul, MN, August 1988.

[23]   Mostow, J., "Design by Derivational Analogy: Issues in the Automated Replay of Design Plans," *Artificial Intelligence,* Elsevier Science Publishers (North-Holland), Vol. 40, No. 1-3, pp. 119-184, September 1989.

[24]   Mostow, J., "Towards Automated Development of Specialized Algorithms for Design Synthesis: Knowledge Compilation as an Approach to Computer-Aided Design," *Research in Engineering Design,* Amherst, MA, Vol. 1, No. 3, 1989.

[25]   Nii, P., "The Blackboard Model of Problem Solving: Part I," *AI Magazine,* Vol. 7, No. 2, pp. 38-53, 1986.

[26]   Nilsson, N., *Principles of Artificial Intelligence (second edition),* Morgan Kaufmann, 1984.

[27]   Ohr, S., *CAE: A Survey of Standards, Trends, and Tools,* John Wiley and Sons, 1990.

[28]   Pentland, A., "ThingWorld: A Multibody Simulation System with Low Computational Complexity," in *Computer-Aided Cooperative Product Development,* Sriram, D., Logcher, R., and Fukuda, S., Ed., Springer-Verlag , pp. 560-583, 1991.

[29]   Requicha, A. and Voelcker, H., "Solid Modeling: Current Status and Research Directions," *Solid Modeling: A Historial Summary and Contemporary Assesment,* pp. 9-24, March 1982.

[30]   Rich, C. and Waters, R. C., Eds., *Readings in Artificial Intelligence and Software Engineering,* Morgan Kaufmann, Los Altos, CA, 1986.

[31]   Rychener, M., Ed., *Expert Systems for Engineering Design,* Academic Press, Inc., Boston, 1988.

[32]   Setliff, D. and Rutenbar, R., "ELF: A Tool for Automatic Synthesis of Custom Physical CAD Software," *Proceedings of the Design Automation Conference,* IEEE, June 1989.

[33]   Sriram, D., Logcher, R., and Fukuda, S., Eds., *Computer-Aided Cooperative Product Development,* Springer Verlag, Inc., 1991.

[34]   Sriram, D., Cheong, K., and Kumar, M. L., "Engineering Design Cycle: A Case Study and Implications for CAE," in *Knowledge Aided Design,* Green, M., Ed., Academic Press, 1992.

[35] Steele, G., *The Definition and Implementation of a Computer Programming Language Based on Constraints,* unpublished Ph.D. Dissertation, Massachusetts Institute of Technology, August 1980.

[36] Stefik, M., "Planning and Meta-Planning (MOLGEN: Part 2)," *Artificial Intelligence 16:2*, pp. 141-169, May 1981.

[37] Stefik, M., "Planning with Constraints (MOLGEN: Part 1)," *Artificial Intelligence 16:2*, pp. 111-140, May 1981.

[38] Steinberg, L., Langrana, N., Mitchell, T., Mostow, J., Tong, C., *A Domain Independent Model of Knowledge-Based Design,* unpublished grant proposal, 1986, [AI/VLSI Project Working Paper No. 33, Rutgers University].

[39] Steinberg, L., "Dimensions for Categorizing Design Tasks," *AAAI Spring 1989 Symposium on AI and Manufacturing,* March 1989, [Available as Rutgers AI/Design Project Working Paper Number 127.].

[40] Steinberg, L. and Ling, R., *A Priori Knowledge of Structure vs. Constraint Propagation: One Fragment of a Science of Design,* unpublished Working paper, March 1990, [Rutgers AI/Design Group Working Paper 164].

[41] Suh, N., *The Principles of Engineering Design,* Oxford University Press, 200 Madison Ave., NY 10016, 1990.

[42] Sussman, G., *A Computer Model of Skill Acquisition,* American-Elsevier, New York, 1975.

[43] Tong, C., "Toward an Engineering Science of Knowledge-Based Design," *Artificial Intelligence in Engineering, special issue on AI in Engineering Design,* Vol. 2, No. 3, pp. 133-166, July 1987.

[44] Tong, C., "A Divide-and-Conquer Approach to Knowledge Compilation," in *Automating Software Design,* Lowry, M. and McCartney, R., Eds., AAAI Press, 1991.

[45] Wilensky, R., *Planning and Understanding,* Addison-Wesley, Mass., 1983.

[46] Williams, B, "Interaction-based Invention: Designing novel devices from first principles," *Proceedings of the Seventh National Conference on Artificial Intelligence (AAAI90),* Boston, MA, pp. 349-356, 1990.

[47] Winograd, T., *Understanding Natural Language,* Academic Press, New York, 1972.

PART VI: KNOWLEDGE ACQUISITION

Chapter 2
A KNOWLEDGE TRANSFER METHODOLOGY IN SELECTION OF MECHANICAL DESIGN COMPONENTS

Manjula B. Waldron and Kenneth J. Waldron

ABSTRACT

The general mechanical design process, particularly at the conceptual level, is far too complex to be amenable to complete simulation by knowledge-based expert systems (KBES) at the present level of that technology. However, once the designer has conceived a preliminary design configuration, he or she quickly moves into problem-solving strategies which require decisions to be made based on prior experience and/or knowledge. These decisions often involve a process of selection of components to be used in the design which is a based, to a high degree, on the individual and cultural experience and knowledge base of the designer. They can be expressed in rules or heuristics to form the knowledge bases of expert systems. The problem is to efficiently formulate the appropriate rules. The use of traditional knowledge acquisition processes, such as schema- or frame-based methods, with protocol analysis, become problematic, since the knowledge engineer may not have the desired technical expertise in preliminary interaction with the designer. The designer, on the other hand, may not be able to voice all the clauses necessary to make the selection. In this paper, the use of CAD tools by the expert mechanical designer, to represent his or her knowledge, is proposed. Mechanical designers are particularly familiar with both CAD tools and decision flow chart representations. The proposed method has the advantage of increasing the expert's role in creating the expert system and making the knowledge engineer's role more efficient. An example from bearing selection is chosen, since these components are widely used in most mechanical systems and provide a paradigm which might be extended to selection of other components such as actuators, sensors, etc.

2.1. INTRODUCTION

In construction of expert systems, a heavy reliance is placed upon the acquisition of accurate knowledge from the experts. This process involves interaction between the knowledge engineer and the domain expert and can be very time consuming if the knowledge domain is ill structured. Hence, methods need to be developed to streamline this process and make it more efficient. In particular, whatever structure is appropriate to the knowledge domain should be used to advantage. This is particularly important as the creation of expert systems moves from the laboratory to the industrial environment [8].

The process of knowledge engineering involves the transfer and transformation of knowledge from the expert to the computer program. Many tools have been developed for the transformation of knowledge using computer programs [7], but the methodology of the knowledge transfer from the expert to a formal form has received relatively little emphasis.

The initial steps of knowledge transfer involve collecting, structuring and identifying knowledge which specifically leads to the solution of the problem. The procedures involve some general steps. That is, the knowledge engineer must interface with the domain expert in such a manner as to minimize interference and allow the expert to solve the problem as he or she would naturally, and must still obtain maximal information. Some aspects of knowledge are specific to a problem domain, and may require tools not necessarily used in other domains. For example, a field such as mechanical design is dependent on visual/spatial cognitive modalities; the natural expression of domain knowledge structure may be by means of these modalities. Hence, if tools which use these modalities are not made available to experts, they may not be able to fully express their cognitive style and thought processes. The result may be a stilted representation of knowledge, or inefficient transfer of knowledge.

Two commonly used methods for knowledge transfer are protocol analysis [5] and directed interviewing [6]. Each method has its strengths and weaknesses, as will be discussed more fully in the next section. However, all the methods described suffer from the common drawback that the interaction with the knowledge engineer may substantially alter the style that the expert uses in solving a specific problem, whether it is done by interview or by collecting a visual-verbal protocol. As Shoenfield cite[Shoenfield83] states, the problem solver's initial belief, as to what is relevant, may subconsciously render large bodies of information inaccessible. Likewise, the individual's reactions to the experimental setting, e.g., fear of failure, projecting a certain image while being audio- or video-taped, may produce social behavior which, too, may alter the style. Ericsson and Simon [5] address the issue of how to make the expert much more comfortable. Nonetheless, the knowledge engineer must be cognizant of this effect. It may be that, at present, this is a natural effect of the process and may be unavoidable. However, these aspects need to be taken into account in the future. This social shyness of mechanical designers was certainly observed in some of the protocols collected by the authors. An interviewer may inadvertently lead the expert into a direction he or she would not normally have considered thereby directing the problem solving rather than capturing the process that he or she would have used. Even more serious is the necessity of translating processes into verbalizations. In a reasoning process which involves extensive use of visualization, demands for verbalization can be very disruptive. This effect has been commented upon by several expert designers with whom we have worked.

In the present paper, these methods, as they relate to conceptual mechanical design, are examined and a process for obtaining knowledge from a mechanical designer is presented. Further, the use of CAD in acquiring this knowledge, by increasing the role of the expert designer in knowledge acquisition and construction of the expert system, is explored. The problem of bearing selection is used as an example to establish the feasibility of the proposed technique.

2.2. SURVEY OF CURRENT METHODS USED IN KNOWLEDGE TRANSFER

The methods used to date fall broadly into several categories, such as verbal think-aloud protocols recorded as the experts solve domain specific problems [5], frame-directed interviews [6], retrospective verbalization and discussion protocols [15, 16], or a combination of these. All methods generally have the same three steps, viz., acquiring the data, analyzing it and, based on analysis, establishing the knowledge required to make decisions which may have been used to solve the problem. A brief review of these methods with a critique of their use in mechanical design follows.

2.2.1. Process Tracing for Initial Knowledge Acquisition

Process tracing is a method of determining the cognitive (thought) processes of experts solving problems in their area of expertise. The procedure is concerned with the type of information used and its function in the decision making process while arriving at a solution. A representative or sample problem is provided to the domain expert and the expert is asked to solve the problem, as he or she normally would, and the protocol is collected. A protocol is defined as a description of the activities (ordered in time) in which a subject engages while performing a task. The power of a protocol lies in the richness of its data. Even though, in themselves, the protocols may be incomplete, they allow a person to see the cognitive processes by which the task is performed, how this process is ordered in time, and what cognitive tools are used in performing the intended task [13].

1. Think-aloud verbal protocols

In this method, the expert is left alone to solve the problem and is asked to speak aloud during the task performance. The expert's solution process is recorded on an audio or video-tape recorder. To confirm the think-aloud procedure, retrospective reporting is sometimes used in which the expert recalls the protocol sequence. Several automatic and semi-automatic protocol analysis programs have been developed [2] with associated linguistic and semantic processors [17].

These programs use problem space and strategy concepts for specific task domains. Examples of applications to thermodynamics, physics, mathematics and simple problem solving tasks like the Tower of Hanoi, are well discussed in Ericsson and Simon's book on protocol analysis [5].

The techniques of protocol analysis require a theoretical framework, along with coding schemes and categories for the types of problems at hand. Verbalization or think-aloud protocol segments need to be encoded semantically in the form of processes which may have produced these segments. Sometimes enough knowledge is available so that, based on the verbal reporting and the task performance, one can predict an acceptable set of sequences to generate a transition network. Usually, though, this is not the case.

This method of generating and analyzing protocols has a distinct advantage of obtaining an expert's solution to the problem, rather than shaping the expert's knowledge to a pre-existing representation system. It minimizes unnecessary interaction between the knowledge engineer and the domain expert. It helps in the collection of knowledge, which is accurate and truly reflects the expert's decision- making process. The knowledge engineer can study the protocol, arrive at the structure at leisure, and interact later with the domain expert for clarification. One limitation of this method, as applied to conceptual mechanical design, is that the expert is simply unable to verbalize, in a satisfactory manner, the spatial visual characteristics of the knowledge and the solution space.

2. Discussion protocols

In this method, two experts discuss and solve the problem, and this discussion is recorded for use in analysis. The discussion provides the knowledge engineer with extra information about the problem from two perspectives and provides decision making alternatives and strategies for resolutions of conflicts in rules. The analysis procedures are similar to those described above in think- aloud protocols. While this method has advantages, such as that of obtaining more general and diverse solutions to the problem, it also has inherent disadvantages. One of the important disadvantages is that the two experts may rely on mutually understood prior information and may communicate this knowledge to each other in many ways other than verbal. In addition, unless the experts share equal expertise, one may overshadow and hence inadvertently intimidate the other, further diminishing the advantage of this method. The knowledge engineer in this type of protocol needs to have a much better understanding of the problem at hand to successfully analyze the protocol and identify the underlying structure of the solution space.

Waldron [15] has provided some further ideas on general protocol analysis which can identify knowledge essential for decision making. He suggests developing a vocabulary and analytical framework for identifying the decision making processes present in the protocols. This structure should focus on the simplicity and the breadth of knowledge rather than on the details as required in the transformation of knowledge into expert systems. He suggests that alternatives, attributes, aspects and appeal could be chosen as classes to which expert's decision making may belong. He further proposes to borrow decision making rules, such as dominance, conjunction, disjunction and maximization, etc., form the decision theory domain to apply to the protocol analysis.

2.2.2. Frame-Directed Interviews

In this method, the knowledge engineer interacts with the domain expert through a formal (directed) or informal interview process. In the informal interview process, the knowledge engineer keeps notes and/or records of the process as it develops. It has the advantage of allowing the knowledge engineer to direct the knowledge acquisition process so it is easy to analyze, but suffers from the disadvantage of being haphazard, inefficient and possibly not a true representation of the expert's cognitive process.

The formal, or directed interview, overcomes these problems. Hall and Bandler [6] propose to apply Minsky's frame theory [11] in the interview process. The frames provide the structure which holds knowledge in slots. Each frame may be theme oriented in which each slot forms the subcomponents of the theme. This method requires the knowledge engineer to be familiar with the terminology and the structure of the problem so as to create the appropriate frames and slots in which to put the data obtained from the expert. The frames must also be created for meta-constructs, i.e. storing information on reasoning about the problem solving process, rather than on steps taken to solve the problem. The data collection sessions are directed by the knowledge engineer, yet the expert must be allowed to display his knowledge in the answers. This requires a clinical psychologist well trained in the interviewing process. The data analysis is also in the form of frames used to build relations between constructs, through the answers obtained to the "how" and "why" questions. This method requires prior planning on the part of the knowledge engineer and a larger interaction time between the expert and the knowledge engineer.

2.3. AN ALTERNATIVE METHOD FOR KNOWLEDGE TRANSFER IN MECHANICAL DESIGN

Mechanical design is a creative process. For instance, the designer's objectives and knowledge bases are more clearly defined than is usual in artistic design. Characteristics of mechanical systems are visually and spatially oriented and the language used is adapted to these perceptual constraints [10]. In the report published from a recent conference on machine intelligence in mechanical design [4], the general consensus was that "The mechanical engineering community working in AI should concentrate on the codification of mechanical information for problem description." Furthermore, for mechanical design the knowledge representation problem is far from being solved.

The conceptual mechanical designer tends to work with simple analytical models and to focus on the principles which are most critical to the viability of the proposed system. He or she draws from a very large and diverse knowledge base, including experience gathered from previous projects, cultural and social knowledge, and knowledge of the technical literature and of available components [9].

To study this process with a view to transferring the knowledge of mechanical designers into a formal characterization form, it was decided that the conceptual designer should have full reign in expressing how he or she will approach representation of a particular design problem.

2.3.1. Bearing selection problem

The problem chosen for representation was selection of bearing types and sizes. Bearings are those elements which control the relative motions of members in a mechanical system and the selection of bearings is ubiquitous in any machine design problem. If there is relative motion or coupling between two parts, the use of bearings of one form or another is mandatory. The bearing type used is dependent on the type of motion, i.e., rotational, or translational, and on motion, load, and lubrication conditions. Bearing selection can occur anywhere in the mechanical design process from conceptual design to detailed design, and, as such, represents a class of component selection problems in mechanical design. The optimization aspects are usually not explicit but may be contained largely in the experiencial knowledge base of the expert. That is, given types are known to be usually optimal in particular types of service. This results in a decision tree based on the expert's heuristics which is amenable to the creation of expert systems.

The selection procedure for bearings is primarily based upon the expected life of the product, the loading and speed condition, and the service environment in which the product functions. The selection procedure, therefore, is a two step process. The type of bearing is usually selected based on the engineer's prior experience, knowledge and any geometric constraints which may be operative. Two or more types may be chosen for comparative analysis prior to finalizing the selection. Depending on the type chosen, the size is determined by analytical procedures. The appropriate equations are often specified by the manufacturer, particularly when dealing with types which are manufactured in standardized sizes. For example, for rolling contact bearings, the designer must transform the actual loading, service conditions and reliability into an equivalent load rating through a series of calculations based on standard formulae. The required bearing sizes are then manually selected by matching the required equivalent load rating with the rated capabilities of each bearing in the catalog.

Bearings are classified into various types, according to their application. Within each major type, as previously mentioned, the primary concern is that the selected bearing must be able to meet service requirements under specific load and speed conditions. Therefore, the load (i.e., its geometry and magnitude) becomes the most important factor in the selection. In the selection procedure, the direction of the load with respect to the shaft axis should be first considered. It can be classified into three types, viz., radial, radial+thrust, and thrust, as shown in Figure 2-1. Depending upon the motion cycle and load magnitude, and on other factors such as reliability, radial space, cost and other service conditions, the choice of the type of bearing can be reduced to a few possible types.

2.3.2. Use of graphics for knowledge representation by expert

To represent this knowledge, the expert felt that a graphical representation was much better suited for explanation of the heuristics than a verbal explanation. He could see the decision flow and felt that drawing on the piece of paper would be much easier and more expeditious than trying to remember and explain every rule and its associated conditions verbally. In view of the use of visual- spatial languages in explaining mechanical design, this was naturally considered to be worth exploring for this problem.

The expert chose MacDraw on the Apple MacIntosh (R) microcomputer as the CAD tool, by which he could graphically represent the kinds of decisions he would make and the types of knowledge he would use in arriving at a solution to the bearing selection problem. The expert developed the selection algorithm directly using this program and expressed the procedure in the form of flow-charts and decision trees. MacDraw, rather than other more sophisticated CAD tools was used because it was a program with which the expert was familiar and allowed easy representation of graphics and text in a form in which the designer was thinking. Further, it allowed the designer to expand his diagram by pasting pages and looking at the multiple pages on the screen as required for review or development. This drawing package is inexpensive, easily available and provided a good test bed.

In order to represent the selection procedure, the expert designer formalized his solution to the problem in a graphical form with selection notes. His initial step in bearing selection was to classify the load geometry. Next he would look at the time history of the projected operation. Next, he would look at the load rating and, based on the type of load from light to heavy, he would make the choice of bearing. This is shown in Figure 2-1. At this point, he thought of another level of constraints which may further dictate the selection. These constraints were the required reliability, geometric constraints, cost, and other applicable service conditions and would make the decision as to the type of bearing he would select (Figure 2-2). If one looks at Figures 2-2 a, b and c more closely, one finds that there is really an other level of constraint added to Figure 2-1, for pure radial or thrust load geometries, while for radial+thrust load geometries, there are no other constraints. The selection, when operating under rotational (radial) or translational (thrust) is further dependent on space and reliability considerations. He felt that cost and type of lubrication which can be used will further determine which bearing would be selected. These constraints were not easy to incorporate into the flow chart and were specified as notes (Figure 2-3). These constraints would have further assisted the expert to decide on which of two alternative bearing types to select.

To look at an example, if the designer knew that the load was radial+thrust, the operating cycle was around 50%, and the load was heavy, then the designer would select a tapered roller bearing. Although the cost of this bearing is higher than that of an angular contact ball bearing, it can support a heavier load for a larger number of motion cycles. Once the bearing type was decided on, such as a hydrostatic or hydrodynamic bearing, he would either use computations based on an analytical model of that bearing type, or go to the appropriate catalog and make the calculations according to the procedure presented there to arrive at the size and stock number of the bearing. In the former case, the bearing would typically be custom manufactured to the designer's specifications. The necessary analytical models and design procedures are described in handbooks and texts on design of machine elements [12].

2.4. CREATING THE EXPERT SYSTEM FROM THE GRAPHICAL REPRESENTATION

This graphical representation, produced by the expert, was successfully converted into a knowledge based expert system using the OPS5 rule based language. The OPS5 programming shell was chosen rather than EMYCIN, since it is more suited to a forward-chaining control flow. The natural structure of the bearing selection process, as represented, is clearly close to forward chaining control flow. OPS5 is a production system programming language which provides a general programming environment and supports the knowledge facts and their relationships as well as an inference mechanism [3]. The main advantage of using OPS5 in the current problem is its stability, efficiency and file action capabilities. The main disadvantage is its lack of well developed user interface. In particular, it lacks editing or explanation facilities, graphics and provides no aids for maintaining test-case libraries. While OPS5 provided the initial language shell for ease of coding and generating the expert system code, it was found that this really was not necessary. The graphical representation has also been successfully used to create an expert system for bearing selection on the IBM PC/AT microcomputer using Golden Common LISP. Currently a graphics shell is being developed on a Xerox 1108 machine to generate the expert system automatically from the graphical representation.

The advantages of this latter environment are many, including cost, speed, portability and small size of the software and system. Figure 2-4 is a sample listing of part of the code in OPS5 and Figure 2-5 shows a portion of an interactive session. When the user is using this system he or she specifies the type of load, its range, reliability, and motion-time history. The program will make the bearing type selection in a manner similar to this expert along with stating the reasons for each step in the process. Further the program would provide ade-

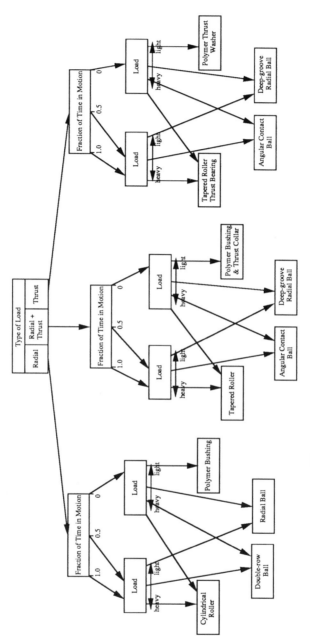

Figure 2-1: Initial Bearing Selection Decisions

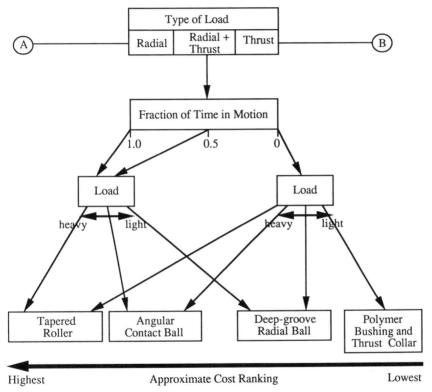

Figure 2-2: Detailed Bearing Selection Decision Tree (a)

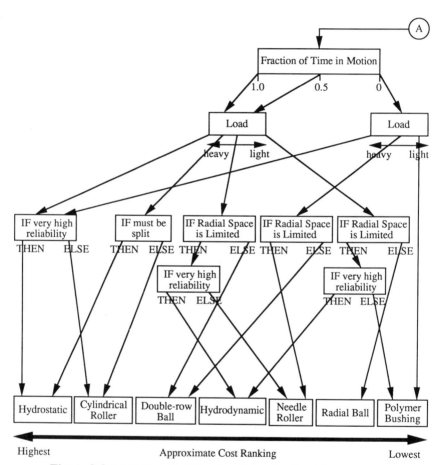

Figure 2-2 (ctd.) Detailed Bearing Selection Decision Tree (b)

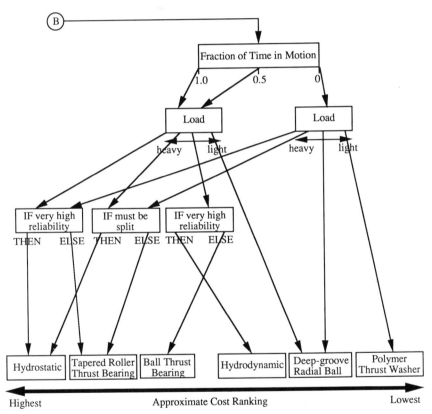

Figure 2-2 (ctd.) Detailed Bearing Selection Decision Tree (c)

Cost:

The bearing types indicated by the heavy boxes are selected from ranges of standard sizes produced in large volumes. The remaining types are custom designed for each application and, hence, their costs are very sensitive to the number manufactured. For example, hydrodynamic shell bearings are very inexpensive in the huge volumes used in automobile engines, but are relatively expensive in small lots.

Lubrication:

External oil pumps are seeded for hydrostatic bearings. Hydrodynamic bearings can be operated either with, or without an external oil pump. Use of a pump increases load ratings. Rolling contact bearing types are normally packed with grease. Polymer bearings are not lubricated.

Thrust and Radial Loading:

Combined thrust and radial loads can always be handled by separate thrust and radial bearings. However, when a combined bearing can be used a more compact arrangement is possible. The combined bearing is also normally less expensive than separate thrust and radial bearings.

Reliability:

Rolling contact bearings have finite life due to fatigue resulting from Hertzian stresses. Their lifetimes can be estimated probablistically. Hydrostatic and hydrodynamic bearings have no contact between the solid elements in continuous operation, and so can have indefinite life.

Figure 2-3: Expert's Notes on Conflict Resolution

quate explanations as to what alternatives may be appropriate based on cost constraints.

2.5. DISCUSSION

It would certainly be naive to assume that such a graphical representation is always possible from every expert or that it is complete. However, for the bearing selection problem, this form provides an adequate knowledge representation and gives a set of rules which guide this expert in making bearing type selection. The notes provide further conditions which were used in coming to a specific selection when more than one choice was possible or when the service conditions, as specified, may not give an optimal solution. The increased availability of CAD tools, and their integration with knowledge system software, make this type of knowledge acquisition method very lucrative for knowledge transfer.

In obtaining knowledge from the expert mechanical designer, it was found that the designer can very clearly and explicitly provide decision conditions for the selection of components by using menu driven graphical tools. These graphical representations can be used directly to create the expert system, which will not only correctly guide the designer to the appropriate components, but will also allow linking to the appropriate databases. When this is achieved the designer has to provide only the design constraints to the software, and the software will provide the appropriate component specifications which can be used in the proposed design. This is currently under development.

The advantages of this technique are numerous. It reduces the burden on the knowledge engineer in creating rules in areas in which his or her expertise is limited. The designer is better able to express the conditions, rules, heuristics, and knowledge in the domain in which his or her expertise lies. The knowledge engineer can use his or her expertise in creating the software which will successfully implement the design specifications.

In this paper, a case is made for providing graphical tools to the design experts so that they may themselves represent their design knowledge and heuristic. IDES [1] is one such attempt. In the IDES system an attempt is made to allow the expert to represent knowledge graphically, which influences adjacent goals in a diagnostic problem. While this system is not directly applicable, nor suited to the present problem, much can be done by using the powerful graphics and symbolic integration available in present day LISP machines, such as the Symbolics 3650. In particular, the use of object oriented languages which allow the designer to graphically represent and correct objects in an efficient manner, is attractive.

```
;* * * * * * * * * * * * * * * * * * * * * * * * * * * * * * * * * * * * * * * * * * *
; ENGLISH VERSION : Rule 5
;                    IF     bearing is " unknown "
;                    and load type is " radial "
;                    and % of time in motion is " >= 0 & < 20 "
;
;                    THEN ask user to specify " load "
;* * * * * * * * * * * * * * * * * * * * * * * * * * * * * * * * * * * * * * * * * * *
(p rule05
    (bearing_selection
      ^bearing {<bearing> unknown}
      ^load_type {<load_type> radial}
      ^motion_ratio {>= 0 <motion_ratio> < 20})
   -->
    (write (crlf) (crlf) What is the load range?)
    (write (crlf) Please specify as requested : : heavy)
    (write (crlf) ------------------------------ : : medium)
    (write (crlf) ------------------------------ : : light)
    (write (crlf) (crlf) LOAD RANGE : :)
    modify 1 ^load (accept))
    (write (crlf) (crlf) (crlf) (crlf)))
;
;* * * * * * * * * * * * * * * * * * * * * * * * * * * * * * * * * * * * * * * * * * *
;ENGLISH VERSION : Rule 6
                     IF     bearing is " unknown "
                     and load type is " radial+thrust "
                     and % of time in motion is " >= 20  & <= 100 "
;                    THEN ask user to specify " load "
;* * * * * * * * * * * * * * * * * * * * * * * * * * * * * * * * * * * * * * * * * * *
(p rule06
    (bearing _selection
      ^bearing {<bearing> unknown}
      ^load_type {<load_type> radial+thrust}
      ^motion_ratio {>= 20 <motion_ratio> <= 100})
   -->
    (write (crlf) (crlf) What is the load range?)
    (write (crlf) Please specify as requested : : heavy
    (write (crlf) ------------------------------ : : medium
    (write (crlf) ------------------------------ : : light)
    (write (crlf) (crlf) LOAD RANGE : :)
    (modify 1 ^load (accept))
    (write (crlf) (crlf) (crlf) (crlf)))
;* * * * * * * * * * * * * * * * * * * * * * * * * * * * * * * * * * * * * * * * * * *
; ENGLISH VERSION : Rule 7
;                    IF     bearing is " unknown "
;                    and load type is " radial+thrust "
;                    and % of time in motion is " >= 0 & < 20 "
;
;                    THEN ask user to specify " load "
;* * * * * * * * * * * * * * * * * * * * * * * * * * * * * * * * * * * * * * * * * * *
(p rule07
    (bearing_selection
      ^bearing {<bearing> unknown}
      ^load_type {<load_type> radial+thrust}
      ^motion_ratio {>= 0 <motion_ratio> < 20})
```

Figure 2-4: Portion of Bearing Selection Code in OPS5

```
-> (make start)
nil
-> (run)

INITIAL BEARING TYPE SELECTION PROGRAM
Please answer all the questions exactly as requested!
Do you want to select bearing? yes or no

ANSWER : : yes

What is the load type?
Please specify exactly as requested : : radial
----------------------------------------- : : radial+thrust
----------------------------------------- : : thrust

LOAD TYPE : : radial

What is the Percentage of Time in Motion?

Please specify between 0 and 100 : : 15

What is the load range?
Please specify as requested : : heavy
------------------------------- : : medium
------------------------------- : : light

LOAD RANGE : : heavy

Does it require very high reliability? yes or no
Please specify exactly as requested!

HIGH RELIABILITY : : yes

CONCLUSION : You have selected
----------------------------------- LOAD TYPE : : radial
---------------------- % of MOTION in TIME : : 15
------------------------------- LOAD RANGE : : heavy
---------------------- HIGH RELIABILITY : :  yes

================ based on these specifications
================ my advice is to choose  :

================ Hydrostatic Bearing

***** COST consideration  :
------- Hydrostatic Bearings are custom designed for each
------- application  ... therefore their costs are very
------- sensitive to the number manufactured.

***** APPROXIMATE C OST RANKING  :
------- the highest of the 7 radial choices

***** LUBRICATION consideration  :
------- Hydrostatic Bearings need external oil pumps whose
------- usage will increase the load ratings.

***** RELIABILITY consideration  :
------- Because of no contact between the solid elements in
------- continuous operation  ... Hydrostatic Bearings can have
------- indefinite life

Please input as requested!
Willl you like to select another bearing? yes  or  quit

ANSWER  : :quit

Thank you for using my services today!

Have a nice day!!!
```

Figure 2-5: Portion of Interactive Session

Future CAD tools need to be designed to accommodate the features discussed in this paper and provide designers with better ways of entering their knowledge and rules. For example, in MacDraw, there was no provision for the designer to easily isolate the segment shown in Figure 2-2 from Figure 2-1. Considerable time was spent reorganizing the entire graph for different pieces. The limited windowing, zoom and reformatting capabilities meant that the expert wasted needless time in making the diagram more readable. In a graphical knowledge acquisition tool for instance, one could provide primitives which are adapted to rapid flow charting. For example, this might include a menu-selected tool to establish a new node, with a framework for naming it and establishing the decision rule which it represents, including provision for appropriate weighting of parameters. Further tools might permit rapid establishment of directional connections between nodes, automatic reformatting to resolve "spaghetti" and permit easy reading of the chart, grouping of sets of interconnected nodes into higher level nodes, etc. Another type of tool would permit attachment of numerical subroutines to nodes to either compute parameters needed for decision purposes or perform design calculations on a selected configuration. A related tool type might permit attachment of a data base of standard component sizes among which a selection must be made.

Further advantages are in automating the creation of the expert system, at least for the type of problems which are similar to the bearing selection problem. The primitives in a knowledge acquisition graphical system, suggested above, would obviously have coding into a set of related rules built into it and the expert need to only specify the decision logic and the system would, in principle, generate the expert system. In the present expert system for bearing selection, once the type of bearing was selected, the user could direct the program to find the appropriate bearing from a data base of catalog information [14]. This involved having the program prompt the user for series of values such as loads, shaft sizes, etc. The software will make the necessary computation in Turbo Pascal and search a data base to find the appropriate catalog number and return to the user with this information, along with the critical calculations. Since the computations are straight-forward, this was easy to accomplish. If flexible tools are created, experts will be able to modify the system as new knowledge becomes available or new heuristics are developed.

2.6. CONCLUSIONS

The method of allowing the expert mechanical designer to graphically express the component selection strategy provides an efficient tool for the transfer of knowledge and heuristics of the knowledge selection process. The efficiency of this method stems from the fact that it allows the expert designer to use tools which are more suited to the visual-spatial mode of thinking which mechanical designers employ. One obtains more information with less interpretation effort.

The graphical representation is further shown to provide potentially an easy means of coding an expert system tool. The expert system designed for selection of bearings in mechanical design, provides complete assistance to the mechanical designer in selecting a bearing fairly quickly. The designer needs only to provide the design constraints. Hence, in itself, it is a powerful CAE tool. This system has the potential to be readily extended to a considerable variety of other mechanical component selection problems, such as selection of actuators, sensors, materials and other components which the designer routinely uses. In fact, the overall structure of the proposed method is applicable to a large subset of mechanical design decisions and forms an informational, if not a cognitive model of this type of selection decision.

The limitations of this technique are also apparent. For example, optimization conflicts are one such limitation. That is, if there is an optimization conflict at the end of the selection process, an expert will probably go to another option quickly by knowing which constraining factors are better optimized than others. Hence, a recursion in the middle is implied and parallel branching may be used to obtain some other piece of information which will be integrated quickly and used to change the design option. The adaptive capabilities of acquiring knowledge and heuristics are lacking in the current expert systems. Without development of better tools, any modification of knowledge or heuristics implies extensive revision of the representation and the corresponding expert system software.

2.7. ACKNOWLEDGMENTS

This work was supported by the NSF Design Theory and Methodology Program Grant #DMC-8516209.

2.8. REFERENCES

[1]   Agogino, A. and Rege, A., "IDES - Influence Diagram Based Expert System," *Proceedings of the 5th International Conference on Mathematical Modelling*, Pergamon Press, 1986.

[2]   Bhasker, R. and Simon, H., "Problem Solving in Semantically Rich Domains," *Cognitive Science,* Vol. 1, pp. 193-215, 1977.

[3]   Brownston, L., Farrell, R., Kant, E., and Martin, N., *Programming Expert Systems in OPS5: An Introduction to Rule-Based Programming,* Addison-Wesley Publishing Company, MA, 1985.

[4]   Cole, J. H., Stoll, H.W. and Parunak, V., "Machine Intelligence in Machine Design," Industrial Technology Institute, Ann Arbor, Michigan, December, 1985.

[5]   Ericsson, K. A. and Simon, H., *Protocol Analysis: Verbal Reports as Data,* MIT Press, Cambridge, MA 02139, 1984.

[6]   Hall, L.O. and Bandler, W., "Relational Knowledge Acquisition," *Proceedings of 2nd IEEE AI Applications Conference,* Miami, pp. 509-513, December 1985.

[7]   Harmon, P. and King, D., *Expert Systems: Artificial Intelligence in Business,* John Wiley and Sons, 1985.

[8]   Hayes-Roth, F., "The Knowledge-Based Expert System: A Tutorial," *IEEE Computer,* Vol. 17, No. 9, pp. 11-28, November 1984.

[9]   Krick, E. V., *An Introduction to Engineering and Engineering Design,* John Wiley, 1969.

[10]   McKim, R. H., *Experiences in Visual Thinking,* Brooks/Cole, Monterey, 1980.

[11]   Minsky, M., "A Framework for Representing Knowledge," in *Psychology of Computer Vision,* Winston P., Ed., McGraw Hill Book Company, pp. 211-277, 1975.

[12]   Mischke, C. R., "Rolling Contact Bearings," in *Standard Handbook of Machine Design* , Shigley, J. R. and Mischke, C. R., Eds., McGraw-Hill Book Company, 1986.

[13]   Ullman, D.G., Stauffer, L.A. and Dietterich, T.G., "Toward Expert CAD," *Computers in Mechancial Engineering,* pp. 56-70, November/December 1987.

[14] Waldron, M.B., Waldron, K.J., and Wang, M., "A Prototype Expert System for Bearing Selection," To Appear in Computers in Mechanical Engineering, 1989.

[15] Waldron, V. R., "Process Tracing as a Method for Initial Knowledge Acquisition," *Proceedings of 2nd IEEE AI Applications Conference*, pp. 661-665, December 1985.

[16] Waldron, M. B. and Waldron, K. J., "A Time Sequence Study of a Complex Mechanical Systems Design," To Appear in Design Studies, 1988.

[17] Waterman, D.A. and Newell, A., "Protocol Analysis as a Task for Artificial Intelligence," *Artificial Intelligence*, Vol. 2, pp. 285-318, 1971.

PART VII: COMMERCIAL APPLICATIONS

Chapter 3
ESKIMO:
AN EXPERT SYSTEM FOR
KODAK INJECTION MOLDING
OPERATIONS

Richard A. Gammons

ABSTRACT

Mold design is a complex task requiring talent and experience. However, once all of the complicated and creative decisions are made concerning how to mold a particular product, there is a great deal of routine knowledge and labor required to finish the design and produce the mold. These tasks are routine, but, none-the-less, require expertise, knowledge, and experience to perform. The ESKIMO project seeks to automate many of these tasks. This will allow the mold design and mold making experts to concentrate their efforts on the creative and challenging aspects of injection mold design and manufacture. In this chapter we describe the mold design task, the principal tool used (The ICAD System^{TM}), and how that tool was applied in the development of ESKIMO.

3.1. INTRODUCTION

Mold Design is a complex task requiring talent and experience. A mold designer needs to make molding decisions which require negotiation with the product designer, knowledge of molding options, knowledge of vendor product offerings, and experience with previous successes and failures. However, once all of the complicated and creative decisions are made concerning how to mold a particular product, there is a great deal of routine knowledge and labor required to finish the design and produce the mold. The routine tasks include:

- refining the design details,
- performing engineering calculations to accurately size the parts,
- defining the commercial parts and raw materials required,
- creating the representation of the design (i.e. CAD models),
- creating dimensioned drawings,
- producing the bills of materials,
- determining what molding machine will accommodate the mold,
- estimating the cost of producing the mold,
- creating process plans for machining the mold components, and
- creating NC toolpaths for the mold components.

These tasks are routine, but, nonetheless, require significant expertise, knowledge, and experience to perform. The ESKIMO project seeks to automate many of these tasks. This will allow the mold design and mold making experts to concentrate their efforts on the creative and challenging aspects of injection mold design and manufacture.

3.2. JUSTIFICATION

The ESKIMO project was initiated to automate some of the mundane processes associated with mold design and manufacture, as well as to provide tools to aid the creative and complex aspects. The ultimate justification for such an effort is in how it affects an organization's ability to produce high quality, rapid turnaround, and cost effective molding services. We expect the most beneficial aspect of the project will be in lead-time reduction thus affecting Kodak's ability to get products to market faster. The quality of products will be improved because designers will have time to investigate more molding options and because certain engineering calculations will be performed in a more rigorous and consistent manner than they might otherwise.

3.3. INJECTION MOLDING BACKGROUND

3.3.1. What is an Injection Mold

An injection mold is designed to produce a particular plastic product. The inverse shape or "impression" of the plastic product is machined into the two halves of a mold insert. These inserts are mounted in a mold base which will pull the inserts apart and eject the product from the insert. The mold base is typically purchased from a mold base vendor and is machined to accommodate the inserts as well as other components. It is also machined to create the runner system, through which plastic flows to the inserts, and cooling lines that will carry cooling water through the mold.

Figures 3-1, 3-2, and 3-3, respectively, show a sample plastic product, the mold insert for the product, and a simplified mold design with space for two mold inserts.

The mold is mounted into a molding machine for making parts. The molding cycle begins with the molding machine delivering hot liquid plastic to the mold. The plastic runs through the runner system, through a gate, and into the cavity formed between the two halves of the inserts. Once the space is filled, the mold machine pulls the two halves of the mold apart and ejector pins are used to push the completed part off the bottom half of the mold base. The mold is then closed up to be ready for the cycle to start over.

In general, the mold itself is made up of a series of plates with screws and pins running through them. Many of the components shown in Figure 3-3 are a part of the mold base as it comes from the vendor. Other components, holes, and features are added to accommodate the particular needs of the product to be molded and the standard practices of the organization.

3.3.2. The Mold Design and Building Process

The individual who designs the plastic product to be molded and the individual who designs the mold are generally different people. The mold designer's involvement usually starts in the latter stages of product design. The mold designer will conceptualize various options for molding the product and will consult with the product designer about the product's geometry as it relates to function and aesthetics. The mold designer may want to add, modify, or subtract features from the product to facilitate molding. An understanding of the aesthetics is necessary so it's known what surfaces of the product can be marked by gates (the point at which the plastic enters the cavity) or by ejector pins (which push on the product leaving marks). With this understanding, a molding concept can be determined.

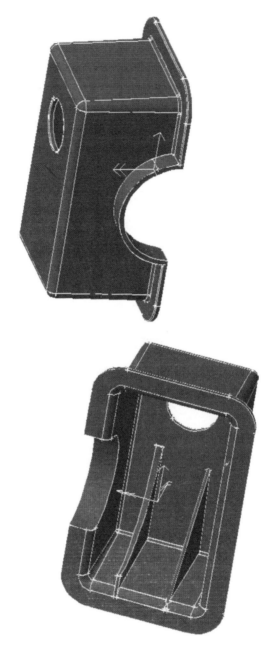

Figure 3-1: A Sample Plastic Product to be Molded

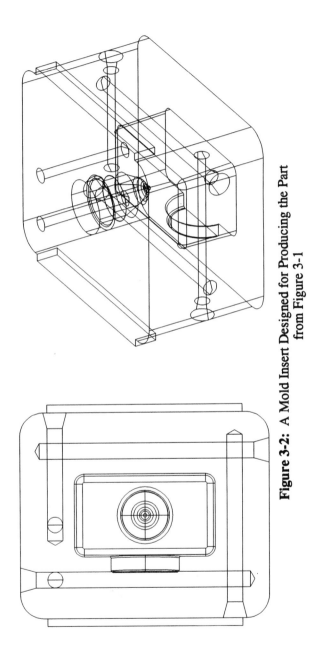

Figure 3-2: A Mold Insert Designed for Producing the Part from Figure 3-1

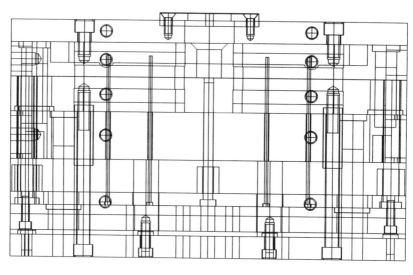

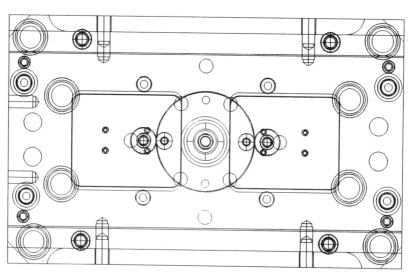

Figure 3-3: A Two Cavity Mold Base. The Base Incorporates Two of the Inserts shown in Figure 3-2

Once the injection mold design is conceptualized, the designer needs to refine the design details: Components are selected from catalogs; geometric calculations are made to correctly lay out the inserts and components; strength and plastic flow calculations are performed to correctly size the parts; and process calculations are made for mold machine selection and setup. The designer will probably run a simulation of how the plastic will fill the mold and make modifications to the mold design based on the results.

As the design details are refined, they have to be documented. Traditionally, this has been done on the drafting board, but, in many companies, the drafting board has been all but retired in favor of CAD systems. Many mold designers will have access to graphic representations of mold bases and catalog components that they can simply retrieve into their CAD files instead of wasting time having to draw these purchased components from scratch. After the geometry is created, it is usually dimensioned so that paper drawings can be produced. The parts lists or "bill-of-materials" are also created as further documentation of the design. At this point, the designer will also have been able to estimate the cost of manufacturing the mold and will be able to compute the unit part cost.

The completed mold design is passed to the mold maker who will machine the parts. In many cases, some of the "design" is actually performed by the mold maker rather than the designer. The designer may have determined where the components belong but the mold maker may be the one who determines the fits and tolerances and other minor, but essential, details.

The inserts are the most complicated parts to manufacture because they contain the impression of the plastic product which might be quite complex. Dozens of other parts require routine machining operations such as hole drilling, pocketing, and trimming. Once the parts are machined, they are assembled into a mold and mounted on a molding machine for producing the plastic product.

3.4. TOOLS USED IN DEVELOPMENT

Playing a minor role in the development of the ESKIMO software were the graphics programming languages within the Unigraphics[2] and ANVIL-5000[3]

---

[2]Unigraphics is a registered trademark of the McDonnell Douglas Corporation, Cypress, CA.

[3]ANVIL-5000 is a registered trademark of Manufacturing and Consulting Services, Scottsdale, AZ.

CAD/CAM systems. (There are two communities at Kodak who are using the ESKIMO software. One uses the ANVIL-5000 system and the other uses the Unigraphics II system.) These programming languages were used to develop interactive programs running in the CAD system that gather the initial set of inputs from the mold designer and from the CAD model of the plastic product.

The rest of the software for ESKIMO has been developed with The ICAD System from ICAD, Inc.[4] The ICAD System is an object-oriented software development environment particularly suited to problems that require physical reasoning along with expert knowledge. The major components of the ICAD System are its language, the compiler, and the ICAD Browser.

3.4.1. The ICAD Design Language

The development environment is built on the LISP programming language. The developer always has access to LISP but does most of the work using the ICAD Design Language (IDL).[5] The following describes some of the features of IDL that make it possible to develop software systems that represent large and complex mechanical systems.[6]

3.4.1.1. Object-oriented definitions

Components of the product are defined as separate "objects". "Objects" are software representations that contain the attributes and behaviors of the thing you are trying to represent. The most intuitive use of objects is to represent real world physical objects such as screws or mold bases, however, objects can also represent non-physical things such as drawing views, manufacturing processes, or operations. The use of objects allows the developer to think about the overall product in terms of relationships between separate parts. The separate parts can be developed individually and assembled later.

---

[4]The ICAD System is a trademark of ICAD, Inc. Cambridge, MA. This document should not be considered as an endorsement of the ICAD product by the author or by Eastman Kodak Company. Another knowledge-based engineering product which provides similar functionality is the Concept Modeller from Wisdom Systems.

[5]The ICAD Design Language and IDL are trademarks of ICAD, Inc.

[6]Portions of the following are excerpted from "Understanding The ICAD System" by Martin R. Wagner of ICAD, Inc.

3.4.1.2. Geometric Parts

IDL provides a set of geometric primitives that contain the attributes and behaviors required of physical objects. Predefined objects include lines, arcs, splines, boxes, cylinders, surfaces, and solids. Boxes, for example, have attributes of length, width, height, location and orientation, as well as volume and specific gravity. Example behaviors for boxes include the ability to display themselves in the ICAD Browser; to produce CAD file descriptions of themselves (e.g. IGES or Unigraphics); to reveal normal vectors from each of their faces; and to be positioned with respect to other objects via English-like constraints (e.g. "in-front" or "top"). The developer can combine these objects to form more complex objects and can augment them with attributes and behaviors relevant to the specific design task.

3.4.1.3. Product structure tree

The product structure tree can be used to represent the way an engineer thinks about the product structure of the mechanical assembly. The product can be divided into its subassemblies, those subassemblies divided further into individual parts, parts are divided into their physical features, and the features are divided into geometric primitives. The tree is not limited in its width or depth. Each level of the product structure tree can be developed separately. This makes it easy to allocate different components of the project among several members of the development team.

3.4.1.4. Mixins

Mixins allow the developer to build complex definitions by "mixing" together simpler definitions. Also, mixins allow the developer to include identical information into more than one object thus avoiding having to repeat the information for each object. Each definition can be made small and easy to understand, which makes the overall product definition very modular and easier to maintain.

3.4.1.5. Symbolic referencing

Symbolic referencing allows the developer to define complex relationships between different parts in the product structure tree. Attributes and behaviors can be accessed by "asking" the object directly. For example, one can access the screw's catalog-number by writing "(the :screw :catalog-number)". This is in contrast to the complex addressing system that one is forced to use when writing large programs in traditional programming languages.

3.4.1.6. Part-whole inheritance

Part-whole inheritance allows the values of engineering attributes in the subparts to be inherited from attributes with the same name higher in the product structure tree. When appropriate, the subpart may override the attribute inherited from its parent. This makes it easy to define engineering values and make them accessible throughout the entire product definition.

3.4.1.7. Relational object manager

The ICAD Relational Object Manager provides relational database tools with which one can make complex data queries and other database operations. ICAD also includes an interface to the Oracle relational database management system.[7]

3.4.2. The ICAD Development Environment

The ICAD Browser (shown in Figure 3-4) is the tool used to create instances of (i.e. "instantiate") and examine the design created by the rules encoded in IDL. It allows the user to flexibly view the product structure tree, to draw the objects in the "Graphics Viewport," and to check the values of attributes associated with the objects in an "Inspector Viewport".

As IDL code is written, it is compiled and the results are tested in the ICAD Browser. Changes are compiled incrementally. This means that the second time an object definition is compiled, only the changes are compiled thus conserving time. Also, there is no "linking" step as is necessary with traditional languages. One can test incremental changes by hitting the "update" button in the ICAD Browser. This saves significant time in the development cycle.

The ICAD System is a "demand-driven" language, meaning that only the computations that are necessary to answer the demands of the user are performed. This is important as one often works on only a small piece of the over-

---

[7]Oracle is a registered trademark of the Oracle Corporation.

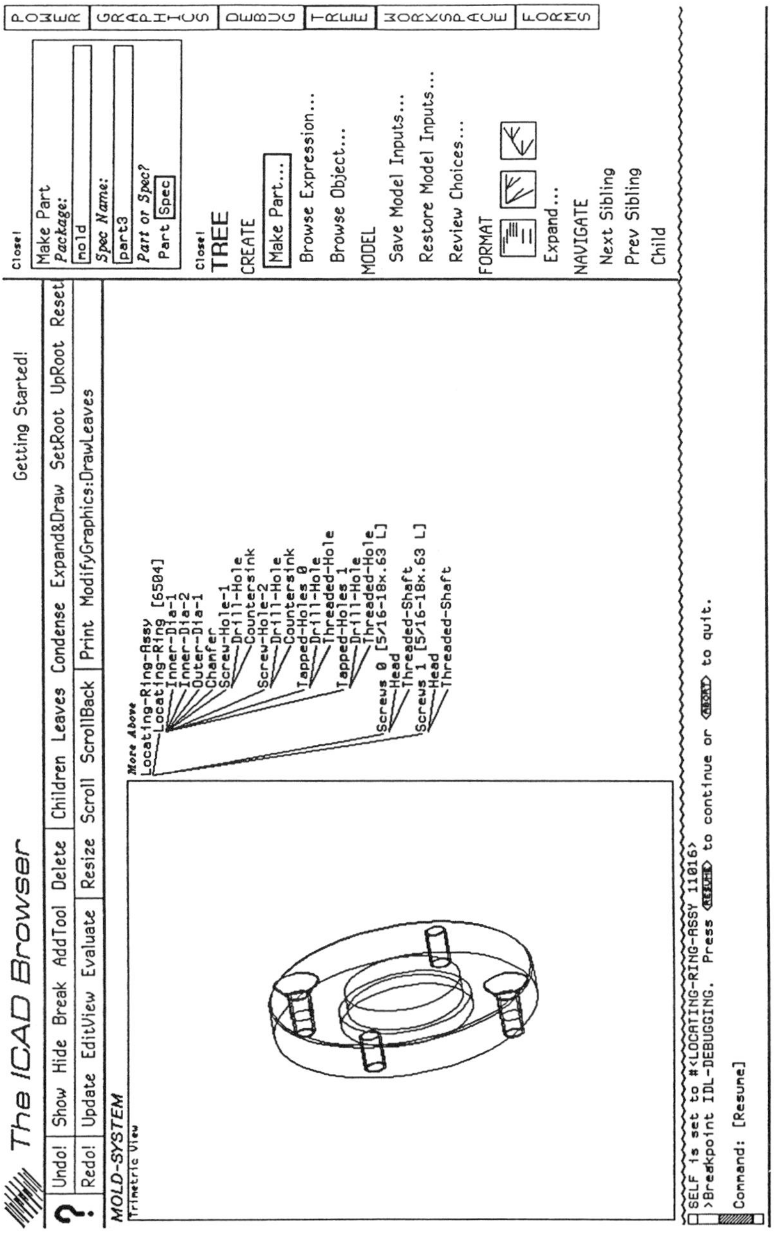

POWER GRAPHICS DEBUG TREE WORKSPACE FORCES

The ICAD Browser

| Undo! | Show | Hide | Break | AddTool | Delete | Children | Leaves | Condense | Expand&Draw | SetRoot | UpRoot | Reset |
| Redo! | Update | EditView | Evaluate | Resize | Scroll | ScrollBack | Print | ModifyGraphics:DrawLeaves |

Getting Started!

MOLD-SYSTEM
Trimetric View

More Above
Locating-Ring-Assy
Locating-Ring [6504]
Inner-Dia-1
Inner-Dia-2
Outer-Dia-1
Chamfer-1
Screw-Hole-1
Drill-Hole
Countersink
Screw-Hole-2
Drill-Hole
Countersink
Tapped-Holes 0
Drill-Hole
Threaded-Hole
Tapped-Holes 1
Drill-Hole
Threaded-Hole
Screws 0 [5/16-18x.63 L]
Head
Threaded-Shaft
Screws 1 [5/16-18x.63 L]
Head
Threaded-Shaft

Close!
Make Part
Package:
mold
Spec Name:
part3
Part or Spec?
Part [Spec]

Close!
TREE
CREATE
Make Part...
Browse Expression...
Browse Object...
MODEL
Save Model Inputs...
Restore Model Inputs...
Review Choices...
FORMAT
Expand...
NAVIGATE
Next Sibling
Prev Sibling
Child

SELF is set to #<LOCATING-RING-ASSY 11016>
>Breakpoint IDL-DEBUGGING. Press <RETURN> to continue or <QUIT> to quit.
Command: [Resume]

Figure 3-4: The ICAD Browser

all design project and has no need to have the rest of the model computed if it has no effect on the task at hand. This is also relevant in a production environment as the user may want to verify only a small, quickly computed, portion of the design before committing to computation of the rest of the model.

Another essential ingredient of the ICAD development environment is the ability to generate various outputs. One example output is shown in Figure 3-3. Figure 3-3 was generated by reading an IGES file into a CAD system. ("IGES", which stands for Initial Graphics Exchange Specification, is a standard file format for exchange of geometric data between CAD systems.) A similar picture could have been generated by creating postscript output and sending it to a postscript printer. ICAD also has specific output capability to various CAD systems including Unigraphics.

In addition to the predefined outputs described here, it is quite simple for the user to define his or her own outputs. The outputs created for ESKIMO will be described later.

3.4.3. An IDL Programming Example

The following example is a simple IDL program that defines an "ejector-pin." The geometry generated by the program is shown in Figure 3-5 and the product structure tree is shown in Figure 3-6.

```
(defpart ejector-pin (box)
 :inputs
 (:pin-length)
 :attributes
 (:width (the :head :diameter)
  :length (the :pin-length)
  :height (the :head :diameter)
  :material '420-steel)
 :parts
 ((head :type cylinder
        :radius 1/4
        :length 3/8
        :position (:front 0.0))
  (shaft :type cylinder
         :radius 1/8
         :length (- (the :length)
                    (the :head :length))
         :position (:rear 0.0))
 ))
```

This simple example demonstrates some new concepts as well as the concepts that were described earlier:

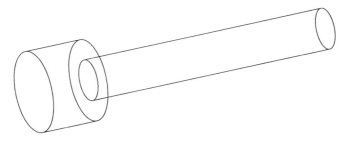

Figure 3-5: The Geometry Created by the IDL specification in
Section 3.4.3 With an Input Value for Pin-length of 2.0.
Note that only the "leaf nodes" are Normally Drawn in the Browser

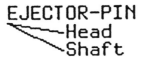

EJECTOR-PIN
Head
Shaft

Figure 3-6: The Product Structure Tree Generated by the IDL Specification
in Section 3.4.3. The "head" and "shaft" are Termed "leaves"
and "ejector-pin" is Referred to as the "parent node"

1. The function "defpart" (which means "define part") is the basic
   mechanism for defining objects in The ICAD System. It contains
   a number of keywords including the ones shown: ":inputs",
   ":attributes", and ":parts".

2. The keyword ":inputs" is used to define attributes of the part that are supplied by the user or the parent defpart. This allows the part to react differently in response to different situations.

3. "Box" is being "mixed in" to the definition of ejector pin thus ejector-pin includes all of the attributes and behaviors of box.

4. The words "box" and "cylinder" are predefined geometric primitives. They were able to draw themselves in the ICAD Browser to produce the drawing shown in Figure 3-5. Typically, the leaves of the tree are used to represent the actual geometry of the part and only the leaves are drawn in the Browser. Geometric shapes associated with parent nodes (such as "box" in the example) are used to help with tasks such as positioning.

5. Symbolic referencing is used in a number of places, the first being in setting the ejector-pin's width to be the same as the head's diameter.

3.5. DESIGN OF THE ESKIMO SYSTEM

3.5.1. ESKIMO'S Product Structure Tree

Figure 3-7 shows the high level structure tree of the ESKIMO system. At the top level is the mold-system. The mold-design is one of the parts of the mold-system as are the mold-machine (the machine into which the mold will fit), the drawings, the bill-of-materials, and the mold-mfg.

The drawings, the bill-of-materials, and the mold-mfg parts are all objects which operate on the mold design and thus are separate from it in the tree. It should be clear, however, that all the information needed to generate the drawings, the bill-of-materials, or the mfg process is contained in the mold design. Only information that is unique to the operation is stored with the operation rather than on the model. For example, all tolerances and data are a part of the model, not the drawings. All catalog numbers, materials list descriptions, and part quantities are a part of the model, not the bill-of-materials. The drawing and bill-of-material parts merely assemble this information for display or output.

Figure 3-8 shows the mold-system tree expanded further. (The symbol "..." indicates that more exists in the product tree that is not currently displayed.) The entire tree is too large to be displayed at once so only certain parts of the tree have been expanded in the figure. At the lowest level, parts are represented with geometric primitives such as lines, arcs, boxes, cylinders, and surfaces. Information describing the parts is stored at whatever level is appropriate. For instance, catalog numbers and materials are stored at the level of the physical part (e.g. "spring" or "screw") rather than at the assembly level or the geometry level.

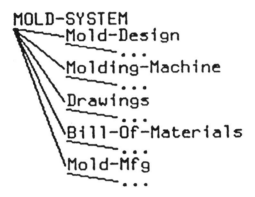

Figure 3-7: The Mold-System Tree

3.5.2. Inputs to the ESKIMO system

The user of the ESKIMO system must provide a number of inputs. First, the product to be molded must be defined. Other inputs involve the major mold design decisions (e.g. hot-runner vs cold-runner, number of cavities, mold-base vendor, etc.). Still other inputs define some of the more detailed mold decisions for which no rules have yet been implemented. Given these inputs, the user expands the design in the ICAD System, looks at the results and decides if the design is appropriate. At this point, he or she can revise the inputs or override some of the decisions made by the system and have ESKIMO revise the design based on the new inputs.

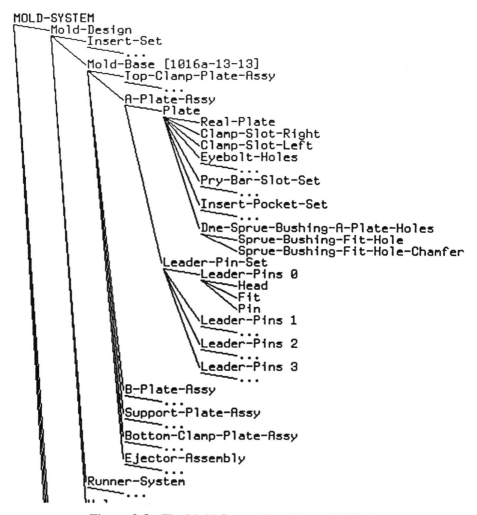

Figure 3-8: The Mold-System Tree Expanded Further

3.5.3. Outputs of the ESKIMO System

Once the design is determined to be ready, the user can generate a number of outputs. Outputs of the ESKIMO system are listed below. Some of these are complete and are being used in the production system and some are still under development.

ESKIMO system outputs:

- Results of calculations (see Figure 3-9)

- 3D wireframe geometry in IGES format (see Figure 3-3), 3D wireframe geometry in a format that can be directly read by the Unigraphics CAD/CAM system (see Figure 3-3 - this would look the same as the IGES file output in this case),

- Solid model output (see Figure 3-10),

- bills-of-materials in various formats (see Figure 3-11), and

- cost estimates and toolpath plans for machining the mold plates (not shown).

The results of calculations can be produced in a number of ways, not the least of which is in the geometry displayed in the ICAD Browser. Another facility provided by ICAD is the ability to create engineering forms using an interactive mechanism. Figure 3-9 shows a sample engineering form used to display some of the non-graphical results of a mold design.

Figure 3-10 shows a shaded image of a solid model that was generated by the ESKIMO system. This was not done with the ICAD Solid Designer (though it could have been). Kodak's development preceded the marketing of the ICAD Solids Designer by ICAD Inc. Without access to this capability, work was done by the developers of ESKIMO to have the injection mold fully represented as a solid model even though no solid modeling type processing is done within the ESKIMO system. This allows the creation of solid model output files that can be read by solid modeling systems.

Mold System

Mold Base Chosen Catalog Number	"1016A-13-13"
Mold Base Price	$1284.50
Mold Machine Chosen	"Van-Dorn-50"
Minimum Insert Wall Thick	1.032
Actual Cavity Thickness	1.377
Parting Line Width	3.625
Parting Line Height	4.375
Mold Plates Machining Cost Estimate	$2376.50

Figure 3-9: An Example Engineering Form Showing the
Results of Calculations

3.6. EXAMPLE OBJECTS OF THE ESKIMO SYSTEM

The various pins used in mold design are a good example of objects. Each pin in the mold is represented by an object which includes all of the information which describes it as well as all of its behaviors. Information which describes the pin includes its catalog number, vendor, diameter, head thickness, length, material, etc. One of the more significant behaviors is the determination of the hole that the pin requires in the mold plates for its operation.

The different types of pins can be described by the same set of attributes. A generic description of a pin is described first and this generic description is mixed in to the description of more specific objects. Figure 3-12 shows a series of pins that are used in the mold base.

The following describes the defparts that were used in the description of the pins:

1. First, a "generic-pin" was defined with all of the information that is common to all pins.

2. The definition of generic-pin was mixed in to the definition of a "catalog-generic-pin." This catalog-generic-pin then has all of the information associated with generic-pin and defines the more

Figure 3-10: Shaded Image of a Solid Model Generated by ESKIMO.
The Image Shows a Subset of the Mold Components
in an Exploded View

MATERIALS LIST

DATE 12-4-1990
TOOL NUMBER PART-DEMO-001   JOB NUMBER 123456
DESIGNER GAMMONS   PHONE UNKNOWN

DET#	HDN	DESCRIPTION	QTY.	MATERIAL
1		A SERIES CUSTOM MOLD BASE	1	
		CAT. #1016A-13-13 420 STAINLESS		
		A PLATE = 1.375		
		B PLATE = 1.375		
		C DIM = 3.0		
		D DIM = 3.99		
		S DIM = 1.3125		
		E DIM = 16.0		
		O DIM = 7/32		
		R DIM = 1/2		
		OR EQUIVALENT		
2		EJECTOR PIN	8	
		EX-9 M-6		
3		GUIDED EJECTOR BUSHING	4	
		CAT #GEB-875		
4		LEADER PIN #5107-GL	4	
5		DME STANDARD LOCATING RING	1	
		CAT #6504 CLAMP TYPE		
6		SPRUE BUSHING B-6601	1	
		O=0.2188 R=0.500		
7		SUPPORT PILLAR #6091	2	
8		EJ-PLT SPRING #MP-36	4	
9		FHSCS 5/16-18 x .625 LG	2	STOCK
10		SHCS 3/8-16 x 1.000 LG	2	STOCK
11	SOFT	STOP COLLAR 1.125 DIA x 1.063	4	CRS
12	SOFT	STOP COLLAR .875 DIA x 1.063	1	CRS

Figure 3-11: A Completed Bill-of-materials Form Generated by ESKIMO

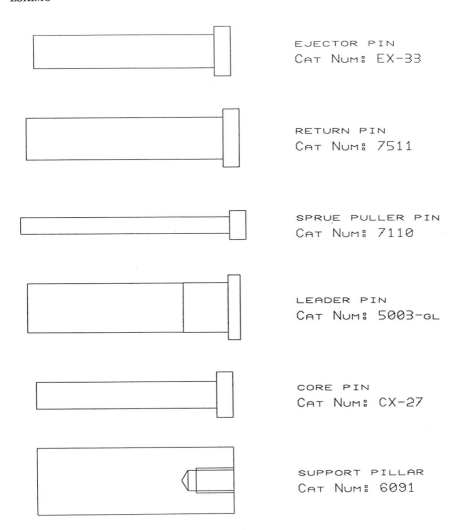

EJECTOR PIN
CAT NUM: EX-33

RETURN PIN
CAT NUM: 7511

SPRUE PULLER PIN
CAT NUM: 7110

LEADER PIN
CAT NUM: 5003-GL

CORE PIN
CAT NUM: CX-27

SUPPORT PILLAR
CAT NUM: 6091

Figure 3-12: Various Pins Used in the Design of the Injection Mold

specific information associated with pins from catalogs. (Conceivably a pin could be custom designed, but, in our case, all pins are catalog-generic-pins.)

3. This catalog-generic-pin is then mixed in to the specific definition of all of the pins shown in Figure 3-9. The specific pin objects add information of how to describe themselves in materials lists as well as other pin specific information. These specific pin objects then contain all of the information and behaviors of generic-pin, catalog-generic-pin, their constituent parts, and the additional information defined within their own defpart.

4. These pin object definitions are used, along with other definitions of parts and part features, engineering calculations, catalog searches, and design rules, to create more complex objects. These more complex objects are further combined to define the injection mold.

3.7. EXPERIENCES

The following lists some of the experiences in the development of ESKIMO:

1. The technology allows the developer to avoid determining the entire design procedure beforehand, as would usually be necessary when developing software with a procedural programming language. One need only be concerned with the definition of objects and the interactions between objects. This has substantial advantages in being able to build systems incrementally and have them be useful to the end user in a partially complete state.

2. It can happen, however, that one ends up with circular reasoning in the interactions between objects. Resolving such situations from the software standpoint is not difficult, especially with the tools that are available to trace the circularity. What is difficult is that it forces you to determine and plan out the design procedure which is something that, before encountering the circularity, you may have been able to avoid doing.

3. Objects should contain all of the information unique to them and should mix in information that is common to a number of objects. Another way to say this is that all data, rules, and behaviors should appear only once. It is quite easy to violate these object-oriented principals when writing code in IDL, but it is important to under-

stand these principals, abide by them, and not be afraid to rewrite code when a more object-oriented approach is realized. This results in much more maintainable and expandable software.

4. A surprise advantage of doing this work has been in the formalism that is being added in performing engineering calculations. We found that many such calculations were being ignored in favor of "rules of thumb" and overdesign. Some were ignored because they so rarely had any effect on the design of the mold, but, only occasionally, would predict an unusual failure mode. The mold designers knew where to find the equations in the text books, but applied them so infrequently that they didn't have much experience with them. By formalizing and computerizing the process, the calculations are always performed and will lead to better quality designs.

5. There is also formalism and consistency being added to general design decisions. We found that there were many things that different designers were doing differently and that these differences were not important to the function of the mold. In many cases, we've been able to agree on standard ways to design the mold and to automate those standards. We expect that this will make a lot of tasks easier including manufacturing the molds.

6. The concept of "concurrent engineering" has received a lot of press lately. In a sense, the ESKIMO program has taken a step towards concurrent engineering in that manufacturing rules have been incorporated into the design process. It was found that some of the design detail of a mold is left unresolved by the designer and only specifically defined by the mold maker. For instance, holes are drawn to accommodate pins or screws, but are not sized precisely by the designer. Sizes and tolerances are left for the mold maker to determine. With ESKIMO, we have consulted with the mold makers and incorporated some of their knowledge into the process to more fully define the mold.

3.8. RESULTS

Development began in March, 1988. Useful versions of the software were available by the Fall of that year. The system is capable of doing roughly 80 percent of the design work on simple two plate molds. Development plans for the remainder of 1990 and 1991 include more refinement on simple molds, the addition of side action, adding commercial parts from additional mold base vendors, improving the user interface, and producing manufacturing cost estimates and toolpath plans. About two man-years have been spent developing the system to date and another man-year of continued development is scheduled for 1991.

The technology employed by the ICAD System is what has made this development possible. It is not likely that the project would have been successful, or would even have been attempted, without ICAD or a similar system.

The team working on ESKIMO consists of Eastman Kodak Company employees Richard Gammons, Kevin Crowe, Steve Trexler, and Mike Blubaugh.

Chapter 4
PRODUCT AND FORMING SEQUENCE DESIGN FOR COLD FORGING

Korhan Sevenler

ABSTRACT

The paper describes a computer-aided design environment for product and process design for cold forging. Commercial tools, DesignView and NEXPERT, were used to develop the design environment. The overall objective of the design environment is to assist the design of cold forged products, the generation of forming sequences, and the design and manufacturing of the dies. The environment consists of a knowledge-based system (KBS), a commercial CAD system, and a finite element method (FEM) based system for process simulation. The paper focuses on the design environment for cold forged product and forming sequence design.

4.1. INTRODUCTION

Cold forging is a broad term that covers various metal forming operations commonly performed in the mass production of axisymmetric parts from steel bars, rods and wire at room temperature. Principle cold forging operations are upsetting (heading), forward extrusion and backward extrusion. These operations are performed in multistage presses to produce parts of various geometric complexity.

Cold forging is becoming increasingly popular due to recent advances made in tool and press design, and in tool materials. In Japan, the total weight of cold forged parts in a car has reached approximately 115 kg [3].

In practice, cold forging requires several "preforming" operations to transform the initial simple billet geometry into a more complex product geometry without any surface or internal defects. After product design, the die designer is faced with the problem of determining the optimum manufacturing sequence (preforming operations) for the product. The design of the manufacturing sequence is a complex and difficult task and so far it has been mostly achieved using experience and trial and error. However, since the die and process development costs are quite high, efforts are made to reduce trial and error based process development.

Several studies have been made to develop knowledge-based systems for forming sequence design in cold forging [1, 4, 8, 10]. However, there is still not a computer system that is widely used by designers for forming sequence design in the cold forging industry.

Cold forging product and sequence design is more an art than a science, involving creativity, intuition and experience. Therefore, instead of trying to develop computer systems to automate this design process, the goal should be to develop computer-aided tools to assist the die designer while they do the design [11].

The following sections describe such a design assistant for product and forming sequence design for cold forging. The system has been developed by Battelle and ERC for Net Shape Manufacturing with the support of several member cold forging companies. The system is developed using two commercial tools; DesignView by Premise Inc. [2] and NEXPERT by Neuron Data [6]. DesignView is used as a two-dimensional interactive graphics environment for product and forming sequence design for cold forging. These designs are then verified by the cold forging Evaluation System which was developed by using NEXPERT. Critical forging stations can also be simulated using DEFORM [7], a Finite Element Method (FEM) based package for metal forming.

4.2. THE PROBLEM: PRODUCT AND FORMING SEQUENCE DESIGN

Cold forging is a metal forming process where metal, at room temperature, is forced to deform plastically into a variety of complex shapes under compressive pressure. Most of the cold forged parts are axisymmetric with relatively small nonaxisymmetric features. Cold forging, as a general term, includes operations such as extrusion, upsetting, coining, ironing and swaging at room temperature. Several forming stations are used successively to form a relatively complex part, starting with a billet of simple shape.

Production of discrete parts via cold forging requires the following major tasks:

1. product design,

2. forming sequence design,

3. die design and manufacture,

4. die and process try-out, and

5. actual production.

Since cold forging can achieve very good tolerances and surface finish, sometimes it is possible to produce the finished product by cold forging. In some cases, the cold forged workpiece may need additional machining to achieve the finished product requirements. Figure 4-1 shows an example, where the designer checks several cold forging limitations to generate the cold formed part geometry from the machine part.

After product design, the next step is the generation of the forming sequence. Several operations are often required to achieve the final cold forged part geometry. The workpiece sometimes needs to be heat treated between these operations. Figure 4-2 shows a possible forming sequence for the cold formed part shown in Figure 4-1. Depending on the designer, alternative sequences are possible for the same cold formed part geometry.

Major savings may be achieved by concurrent product and forming sequence design in cold forging. Minor modifications of the product geometry may make cold forging of the part possible, and reduce or eliminate secondary machining processes (Figure 4-1). Therefore, it is important that the computer system supports the design of the cold forged product and the forming sequence simultaneously. The system must support product design requirements such as interactive graphics and dimensioning.

In the early stages of forming sequence design, the designer draws rough sketches of the preforms for each station. He, then, calculates the volumes and tries to establish the accurate workpiece dimensions (Figure 4-2). The interactive graphics system should support the creation of the initial concepts easily and help to determine the workpiece volume. The designer may then want to change some of the dimensions to see the changes in other dimensions while keeping the workpiece volume constant. The interactive graphics system should be customized for such what if analysis for cold forging.

A knowledge-based system (KBS) integrated with the interactive graphics system can be used for sequence generation and verification. The KBS is used mainly for the verification of the sequences generated by the help of the design system. For some product geometries, the KBS can generate the forming sequence automatically [10]. The KBS should not only have cold forging rules and guidelines such as maximum reduction ratios for forward and backward extrusion, but it should also include simple analysis such as the prediction of average forging pressures and extrusion defects.

FEM simulation of the metal flow and the tool deflection may be required for some critical processes in the forming sequences. Therefore, the computer environment should include such an FEM program. The best candidates for this task is a computer program, called, DEFORM, an FEM program developed by the Battelle Columbus Division for forging simulation. DEFORM was recently used to simulate metal flow and tool deflection during cold extrusion process [7].

MACHINED PART   **COLD FORMED PART**

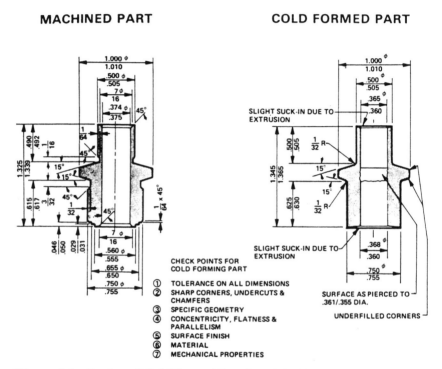

Figure 4-1: Design of Cold Formed Part from Machined Part Geometry [4]

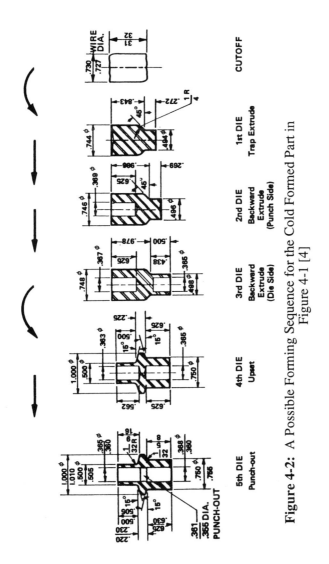

Figure 4-2: A Possible Forming Sequence for the Cold Formed Part in Figure 4-1 [4]

4.3. COMMERCIAL TOOLS USED

Two commercially available tools, DesignView by Premise Inc. and NEX-PERT by Neuron Data were used in developing this system as described in Section 4.4. This section gives a brief overview of these commercial packages.

4.3.1. DesignView™

Conventional CAD tools help engineers capture geometry to show how the design will look. Traditionally, these tools do not offer the ability to incorporate the functional and manufacturing requirements of a product into the CAD model. DesignView by Premise Inc. is one of the best packages in the market that deals with this shortcoming with its dimension-driven variational geometry approach.

4.3.1.1. Core technology

DesignView's core technology is a set of algorithms and object-oriented database structures which unifies variational geometry and mathematics [9]. Variational geometry, one element of DesignView's technology, is a constraint-driven geometric modeling technique. Constraints may be dimensions or other geometric relationships such as tangencies and co-linearities. When designer changes a dimension value or adds a new constraint, a constraint solver computes any changes in the model and then redraws the entire geometric model to reflect those changes.

Variational geometry implies that constraints are non-directed and solved simultaneously. For example, in forming sequence design, the designer can set the volume of the billet equal to the volume of the forged part, and he sets the billet diameter to a known wire size to compute billet length, or he can set the billet length to a desired height to compute the billet diameter. Other examples of the use of non-directed constraints are shown in the next section. The key difference between variational geometry and parametric geometry is that in a variational system the user can solve for any parameter in the model at any time. This means that, unlike a parametric system, a variational system can back solve.

The unification of geometry and mathematics in a single model is another key strength of Premise's technology [9]. In the same way that geometric constraints and dimensions control the *form* of a component or system, mathematics provides a means of specifying *function*. For example, with non-directed mathematics, a mathematical model of a table saw that computes the power required to a cut a given wood thickness, can also compute the maximum wood thickness that can be cut for a given power setting without changing the mathematics.

Unifying variational geometry and mathematics is a requirement for concurrent product and process design. Applications in product and process design involve interactions between mathematically-constrained functional matrices (flow stress, reduction ratio, forging pressure) and geometrically-constrained form. DesignView is one of the first packages that unifies geometry and mathematics into one fully integrated modeling system, not two separate models (one for mathematics and one for geometry) linked together [9].

An important feature of DesignView is that it can communicate with other MS Windows applications using Dynamic Data Exchange (DDE) protocol. An example of such a communication is described in Section 4.4.

4.3.1.2. Key analysis features

DesignView's equation solver can solve an unlimited number of simultaneous non-linear equations [2]. The user can also include inequalities. It has several functions for computing mass properties such as area, null axes, centroid, polar moment, Ixx, Iyy and Ixy. It is also capable of mass property synthesis, that is, it can define the geometry for the desired mass property.

DesignView has several built in functions for trigonometric, hyperbolic, logarithmic and exponential functions. It also has user-defined functions where the user can define a function and then use it in several equations in the same worksheet. DesignView also supports a form of If-then-else function which is useful in describing the design model. For example, If-then-else function can be used together with the Beep and Warn functions to develop worksheets which will later warn the user for the limits of the design.

DesignView can import and export IGES and DXF files. This is important since in some cases the initial design in DesignView may have to be exported to CAD system for other tasks. For example, once the conceptual design of the product is completed, it can be transferred to a CAD system for the design of the dies and then the generation of cutter paths for the EDM machine.

DesignView has other miscellaneous functions such as animation and automated iteration of variable value which can be useful for tasks such as linkage tolerance analysis.

4.3.1.3. Key drawing features

The drawing entities of DesignView include lines, arcs, circles, points, cubic and quadratic splines. The designer can use the construction elements such as horizontal line, perpendicular line, angular line, parallel line, crosshair lines, construction circles for defining their model. Construction lines provide suggestive feedback during drawing [2]. Geometric constraints such as go through point, fixed point, make tangent, parallel, and perpendicular, are used to constrain the geometry wtih the equations as described in Section 4.3.1.1.

DesignView supports cartesian and polar coordinate systems in both absolute and relative coordinates. The coordinate originate can be moved to the desired location on the worksheet. To speed up the creation of the design, it supports a grid with separate X and Y spacing, and adjustable angle. The user can choose the linear units as inches, feet, millimeters, centimeters, or meters, and the angular units as degrees, or radians.

DesignView has other miscellaneous drawing features such as cross-hatching and associative fillets and chamfers.

4.3.2. NEXPERT *Object*™

NEXPERT *Object* by Neuron Data is a hybrid rule and object base expert system building tool. A major strong point of NEXPERT is that it is designed to allow easy integration to existing computing environments.

4.3.2.1. Knowledge representation

NEXPERT uses two sets of structures for knowledge representations: rules and objects.

Rules

The rules are the elementary chunks of knowledge processed by NEXPERT. A rule is a mode of representation of dynamic chunks of knowledge, linking facts or observations to assertions or actions [5].

NEXPERT rules are not necessarily defined as forward or backward. Depending on the inference during a specific session, a given rule may be processed in either forward or backward direction. Neuron Data calls this format Augmented Rule Format (ARF) [6].

All elements of a rule are objects or classes except the operators, constants and some types of arguments. Rules also have attributes. An important rule attribute is the rule category. The rule category attribute is an important feature which makes it possible to solve conflicts in a declarative manner and restrict the processing modes of the rule.

Objects

An object in NEXPERT is an instance or prototype, a structured description of an item corresponding to a precise reality [5]. In NEXPERT, objects can be dynamically created by the rules, and the relationships between the existing objects can be modified through inference.

An object has a set of properties to describe its qualities in symbolic or numerical terms. An object may belong to several classes and it may also have subobjects which may constitute its components. For example, one may have a class named "radially extruded parts." Some of its subclasses will then be "spiders", "inner races" and "pinions." A specific pinion may then be an object belonging to the "pinions" class. This object may then have subobjects called "tooth" and "through hole."

For each property of an object or class, there is a series of parameters which may be customized, such as inheritance relationships or sources of information. NEXPERT supports multiple inheritance which means that an object can have two or more parents.

4.3.2.2. Knowledge processing

Knowledge processing in NEXPERT is controlled by the Agenda mechanism. This is the mechanism by which events are scheduled to happen during knowledge processing. NEXPERT has a rich set of control mechanisms in the Agenda: Backward Chaining, Gates, Right Hand Side Actions and Contexts. By combining these mechanisms, one can have complicated control strategy during knowledge processing. These different control mechanisms can be dynamically turned on/off during a session.

An interesting feature of NEXPERT is that the knowledge can be represented as a set of knowledge islands using the Contexts. During knowledge processing, the control can *focus its attention* on one such knowledge island.

4.3.3. Application Programming Interface

A major advantage of NEXPERT is its open architecture. NEXPERT can communicate with other processes such as triggering external actions when it reaches some conclusions. It can also receive events from the external environment and modify its internal agenda to process these events [5]. The core AI library of NEXPERT can also be embedded in an application in which case these communications become much faster. It also has built in functions for database access and supports several commercial databases.

An example of NEXPERT communicating with another application, DesignView, is described in the next section.

4.4. IMPLEMENTATION

Figure 4-3 illustrates the overall environment for concurrent product and process design for cold forging. There are three major components of the environment:

1. a design assistant, called FORMEX,

2. a CAD System, and

3. an FEM based program, called DEFORM.

FORMEX is a computer program developed using DesignView and NEXPERT for the initial product and forming sequence design, and evaluation. The forming sequences designed and evaluated by FORMEX are then transferred to a commercial CAD system for die design and manufacture (Figure 4-3). The designer can also simulate the metal flow for the critical operations in the sequence using an FEM based program called DEFORM [7]. Process simulation with DEFORM allows the prediction and avoidance of internal and surface cracks.

The rest of the section describes the two major components of FORMEX (Figure 4-3):

1. the Design System for initial product and forming sequence design, and

2. the Evaluation System to verify the results of the forming sequences.

4.4.1. Design System

The design environment is developed using DesignView, a variational geometry package described in Section 4.3.1. The environment supports cold forging product design as well as forming sequence generation.

For cold forging product design, old product geometries are kept in a product database from where they are retrieved when required. The Design System is a specifically important tool for the conceptual design of new products and the parametric design of existing products. During product design, cold formability of the new product is simultaneously checked by generating alternative forming sequences and verifying them.

For forming sequence design, the user starts with the product geometry. The first step is to check the forming sequences library (Figure 4-3) for an existing sequence for this part family. If there is such a sequence, it is retrieved and displayed on the monitor. Figure 4-4 shows such a sequence retrieved from the sequence library. The designer, then, has several interactive design tools to modify the sequence for the given product. For example, he can define the upset volume to be constant, and try different dimensions for the upsetting station. The designer can create several alternative sequences and then verify them with the evaluation system described in Section 4.4.2.

If there is not a forming sequence in the library with which to start, the designer can use the operations that are stored in the cold forging operations library (Figure 4-3) to establish an initial sequence from scratch. The designer starts with the final product geometry and may design backward toward the billet, or he first finds the billet dimensions, and then designs the preforms toward the final product. In each case, he will use the template operations that are in the operations library.

The operations library was developed as a set of DesignView worksheets. There is a separate worksheet for each cold forging operation in the library. The worksheets were later organized as cards in a DesignView Application Library. The Application Library permits the designer to search through different operations before selecting the desired operation.

There are two types of operations in the cold forging operations library :

1. Specific cold forging operations.

2. General (basic) cold forging operations.

Following are some examples of specific cold forging operations and how they can be used to generate forming sequences. The general (basic) cold forging operation, such as forward extrusion, backward extrusion and upsetting, are used similarly and explained in [11].

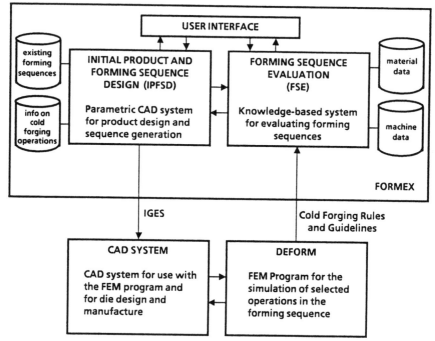

Figure 4-3: Computer-Aided Design Environment for Cold Forging
Product and Process Design

Specific operations are actual operations that are extracted from previous forming sequences. An example of a specific operation is shown in Figure 5. This is a case of simultaneous forward and backward extrusion. The workpiece on the right is the preform before forging, and on the left is the workpiece after forging. Only half of the cross section is drawn to simplify the worksheets. Since these are rotationally symmetric products, the half is adequate to represent the product.

In the worksheet shown in Figure 4-5, the designer may change the dimensions to create different workpiece geometries. For example, he can change the depth of backward extrusion by changing the value of the height fh1. If he changes the value of fh1, given the volume of the workpiece is constant during forging, the new dimensions are calculated automatically. Figure 4-6 shows the new workpiece after the value of fh1 is changed from 1.040 *in* to 1.300 *in* by simply entering the new height on the screen. The corresponding change on the preform (on the right) are set by the system. Similarly, the designer may change

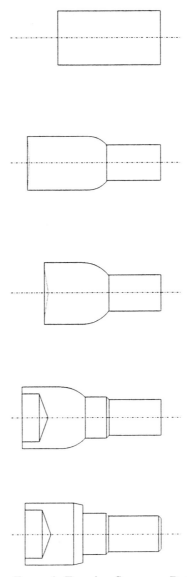

Figure 4-4: An Example Forming Sequence Retrieved from the
Forming Sequences Library

(Dimensions are in the library and, for clarity, are not displayed)

the radius fr1 from 0.495 *in* to 0.600 *in* and new workpiece geometries are calculated automatically as shown in Figure 4-7.

Another example - forging operation, piercing and ironing - is shown in Figure 4-8. In this case, let's assume that the designer wants a smaller part with a smaller inner radius and shorter height. If the inner radius, fr3, is changed to 23.0 mm, and fh3 is changed to 230.0 mm, the new product and preform geometries are calculated and displayed as shown in Figure 4-9.

The template operations in the cold forging library are used one after another to find a possible sequence. The designer can also include a different operation in the sequence if he can not find an appropriate one in the library. Several alternative forming sequences can be designed using the Design System. The next step is to verify the sequence using the Evaluation System (Figure 4-3).

4.4.2. Evaluation System

The Evaluation System is a knowledge-based system for verifying forming sequences designed in the Design System or elsewhere (Figure 4-3). The Evaluation System is developed using NEXPERT which was described in Section 4.3.2.

The Evaluation System accesses two databases, one for material data and another for machine data (Figure 4-3). The material database stores the material specifications such as the extrusion and upsetting limits and material constants for flow stress calculations. The machine database stores the machine specifications such as the number of stations, cutoff diameter, minimum and maximum cutoff and kickout lengths. These material and machine specifications are retrieved from the database whenever they are required during a forming sequence evaluation.

The required geometric variables are received from the DesignView worksheet using the MS Windows Dynamic Data Exchange (DDE) links. These values are assigned to the corresponding data slots in NEXPERT, which are then used in NEXPERT rules.

The Evaluation System verifies the forming sequence as a whole or focuses on a desired station. When it focuses on a specific station, it checks several constraints for the assigned operation. For example, there are three major functions for the open die extrusion operation as follows:

1. check extrusion reduction ratio,

2. estimate optimum die angle, and

3. compute extrusion load.

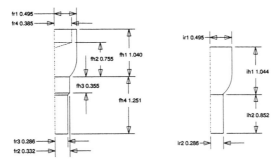

Figure 4-5: An Example of a Cold Forging Operation from the Operations Library

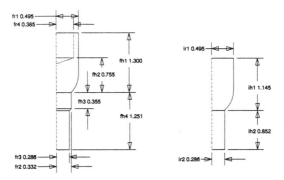

Figure 4-6: The Cold Forging Operation in Figure 4-5, after Height fh1 is Changed

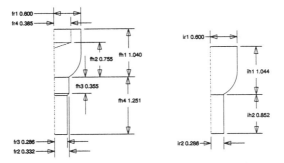

Figure 4-7: The Cold Forging Operation in Figure 4-5, after Radius fr1 is Changed

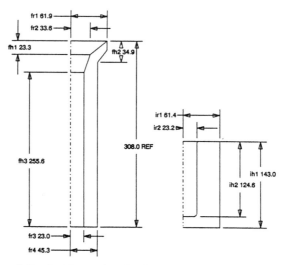

Figure 4-8: An Example of a Cold Forging Operation from the
Operations Library

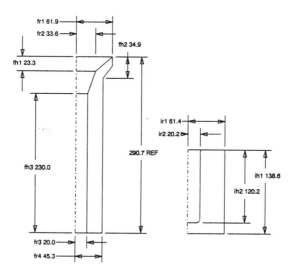

Figure 4-9: The Cold Forging Operation in Figure 4-8,
after Inner Radius fr3 and Height fh3 are Changed

The system compares the reduction ratio with the suggested limit for the given material and informs the user if the limit is exceeded. It suggest the optimum die angle for minimum force requirements and can also predict the required extrusion load with reasonable accuracy. Similar functions exits for other single station operations such as trapped die forward extrusion, backward extrusion and upsetting.

The user may specify a forging machine for the suggested forming sequence. The system will then retrieve the machine specifications from the machine database and check whether the sequence violates any machine requirements.

The Evaluation System not only has common cold forging rules, but it can also perform simple analysis. For example, it can predict the average pressure and forging loads with well-tested formulae, mostly base on the slab method analysis. Several criteria for predicting chevron and external defects in multistage forward extrusion have been studied. Methods for preventing these defects are currently being included into the Evaluation System.

4.5. SUMMARY AND CONCLUSION

A computer-aided design environment for product and forming sequence design for cold forging is discussed. The environment, called FORMEX, runs on IBM PC's and compatibles with MS Windows. Two major components of the environment are: (1) an interactive Design System customized for cold forging product and process design, and (2) a knowledge-based Evaluation System for the verification of these forming sequences.

The forming sequences are exported from the system in standard IGES and DXF formats. This makes it easier to import the forming sequences in commercial CAD systems for (1) further verifications of the sequence using FEM programs, and (2) designing and manufacturing the dies (Figure 4-3).

Initially the system was being developed using Prolog on a VAX machine [10]. When the decision was made to use a commercial tool, it was discovered that no single tool can support the requirements. It was decided to use DesignView and NEXPERT together which turned out to be a good combination for this application.

Finally, the objective of the project is not to develop a computer program that can automatically generate forming sequences, but to develop a computer-aided design tool that can assist the die designers in forming sequence design. Establishing forming sequences is a creative task, and the die designers will do a much better job than computer programs in many more years to come.

4.6. ACKNOWLEDGMENTS

The author acknowledges the support of ERC for Net Shape Manufacturing and several member companies in cold forging. He also thanks National Machinery Company of Tiffin, Ohio for providing valuable information and suggestions.

4.7. REFERENCES

[1]   Bariani, P. and Knight, W. A., "Computer Aided Cold Forging Process Design: A Knowledge-Based System Approach to Forming Sequence Generation," *Ann. CIRP*, Vol. 37, No. 1, pp. 243-246, 1988.

[2]   *DesignView User's Manual*, Premise Inc., Three Cambridge Center, Cambridge, MA, 1990.

[3]   Kudo, H. and Takahashi, "Extrusion Technology in the Japanese Automotive Industry," *Cold Forging - 8th Int. Congress*, VDI Berichte 810, VDI Verlag, Dusseldorf, Germany, pp. 19-36, 1990.

[4]   Lange, K. and Du, G. H., "A Formal Approach to Designing Forming Sequences for Cold Forging," *Proc. of 17th NAMRC*, pp. 17-22, 1988.

[5]   *NEXPERT Object Description*, Neuron Data Inc., 1 56 University Avenue, Palo Alto, 1989.

[6]   *NEXPERT Object Fundamentals*, Neuron Data Inc., 1 56 University Avenue, Palo Alto, 1989.

[7]   Oh, S. I., Wu, W. T., Tang, J. P. and Vedhanayagam, A., " Simulation of Metal Flow and Tool Deflection during Cold Extrusion Process," *Proc. of ASME Winter Annual Meeting*, 1990.

[8]   Osakada, K., Kado, T., and Yang, G. B., "Application of AI Technique to Process Planning of Cold Forging," *Ann. CIRP*, Vol. 37, No. 1, pp. 239-242, 1988.

[9]   *Technology Background*, Premise Inc., Three Cambridge Center, Cambridge, MA, 1990.

[10]   Sevenler, K., Raghupathi, P. S., Altan, T., and Miller. R. A.,
        "Knowledge-Based Approach to Forming Sequence Design for Cold
        Forging," in *Knowledge-Based Expert Systems for Manufacturing*,
        ASME PED-Vol.24, pp. 299-310, 1986.

[11]   Sevenler, K. and Altan, T., "Computer Applications in Cold Forging -
        Determination of the Processing Sequence and Die Design," *Cold Forging - 8th Int. Congress*, VDI Berichte 810, VDI Verlag, Dusseldorf, Germany, pp. 209-221, 1990.

Chapter 5
AN OPEN-ARCHITECTURE APPROACH
TO KNOWLEDGE-BASED CAD

David J. Mishelevich, Matti Katajämaki, Tapio Karras,
Alan Axworthy, Hannu Lehtimäki
Asko Riitahuhta, and Raymond E. Levitt

ABSTRACT

The Integrated Boiler Plant Design (IBPD) system implemented at Tampella in Finland uses Design^{++}, which is an object-oriented knowledge-based CAD system built over Intellicorp's KEE. Design^{++} supports concurrent engineering and provides open-architecture interfaces to CAD packages, relational databases, external engineering-analysis packages, and desktop publishing systems. In addition it also provides support for rule-based inferencing. The IBPD system utilizes all these features of Design^{++}.

5.1. INTRODUCTION

Design^{++tm} is an open-architecture Knowledge-Based Computer-Aided Design (KB-CAD) tool. Design^{++tm} implements its symbolic-modeling capabilities in the object system of IntelliCorp's Knowledge Engineering Environment (KEE^{TM}), provides for building applications as product structures constructed from assemblies and parts contained in project libraries, incorporates design rules that are deterministic engineering expressions, and provides the "glue" CAD systems, external data sources such as relational databases or engineering-analysis packages, and desktop publishing systems. In this chapter, after a brief discussion of KB-CAD, a Design^{++tm} application, the Integrated Boiler Plant Design (IBPD) system of the firm Tampella in Finland is described, including achieved benefits. This is followed by descriptions of Design^{++tm} it-

125

self and its support of concurrent engineering, additional application examples, training, knowledge publishing, and future growth plans.

5.2. KNOWLEDGE-BASED COMPUTER-AIDED DESIGN

In conventional CAD, components, such as pumps, can be composed of associated lines and circles, that can be called out of a library. Whereas today CAD mainly is a drafting tool, knowledge-based CAD (KB CAD) leverages engineering knowledge. KB-CAD adds the following abilities to conventional CAD:

- Modularize the design into projects, libraries, assemblies, subassemblies, and parts,

- Contain functional information about a given object (e.g., weight, capacity, cost),

- Use a generic product-structure definition and rules to determine which components should be inserted into a new instance of the product,

- Use smart components (such as structural beams that know how to size themselves in the context of known loadings and supports),

- Maintain relationships among components, and

- Automatically generate reports, including bills of materials, describing the system.

KB-CAD tools, in general, can be effectively utilized for the semi-custom design of almost anything. The idea is to configure systems out of available components that are the building blocks. Using KB-CAD tools engineers can design by making selections from libraries of assemblies and parts that know how to size or connect themselves. The resultant product structures are easily understood. Modifications can quickly be accomplished to update and expand the product being designed.

KB-CAD tools can be utilized in various stages of the design process, e.g.,

- Pre-sales product configuration,

- Conceptual design,

- Cost estimation (with assemblies and/or parts individually estimated to an appropriate level of detail),

- Product design at the schematic level, and

- Product design at the detail drawing level.

The earlier in the evolutionary stages of a product life cycle that KB-CAD is applied, the more leverage is obtained. The earlier stages are the most knowledge intensive. Conceptual-design decisions can make enormous impacts on the later detailed stages. In addition, mistakes or non-optimal decisions caught early in the overall process can save large quantities of time, stress, and money.

In the next section, we will describe an application of a KB-CAD tool - Design^{++TM} - for boiler plant design.

5.3. TAMPELLA INTEGRATED BOILER PLANT DESIGN SYSTEM

The Integrated Boiler Plant Design (IBPD) system was the first Design^{++} application and provided the genesis of the tool. Thus, Design^{++} was considerably influenced by an industrial application.

5.3.1. The Application

The rationale for the Tampella IBPD system was that competition within the heavy engineering industry has become very intense because of increasing saturation of the market for boilers and boiler plants. One way to improve competitiveness is to improve efficiency of engineering design. A critical element of decreasing cost is getting plants into production faster and thus generating revenues and profits earlier. Improving delivery times for plants (say from 25 to 20 months) can have a dramatic impact on everything from the cost of building to the amount of interest paid on construction funds.

The part of the power plant which generates high-pressure steam utilizing different fuels is called the boiler plant. Design of the boiler plant is driven by first the high-pressure steam requirement and second the characteristics of the fuel to be burned. Tampella had significant experience with their Computervision CAD systems, and while excellent results were obtained, the company determined that they had reached a wall in terms of additional progress. The parametric programs in the CAD system were typically becoming 20,000 to 30,000 lines in length. Maintenance of programs of this size, particularly in light of changes required in the rules governing the design, was overwhelming. This catalyzed the birth of the IBPD knowledge-systems approach.

Among the objectives of the Integrated Boiler Plant Design were:

- placing more of the Tampella knowledge about boiler design in an organized and readily available fashion prior to the initiation of the design of a specific plant, and

- improving conceptual design by making the constraints inherent in the various disciplines involved in the design available to all the other disciplines.

The enhanced body of Tampella boiler knowledge is called the Tampella Super Boiler. The development of IBPD began in July of 1986 and it went into initial production in April of 1987. The application has undergone continuous expansion and refinement since.

5.3.2. Complexity of the Design Process

Boiler plant design is both very complex and repetitive. Table 5-1 organizes considerations of boiler plant design into four categories [5]. The flow of data in the design of a boiler plant is shown in Figure 5-1. This and the rest of the figures in this chapter relating to IBPD are either reproduced from or modifications of figures appearing in [5].

5.3.3. Emphasis on Preliminary Design

Tampella divides the design process into three stages: proposal design, preliminary boiler plant design, and detailed design. In this chapter, the phrase preliminary design is used to cover stages one and two, and the term implementation design covers stages two and three. It was recognized at the initiation of the IBPD project that the most leverage could be obtained during the preliminary or conceptual design stage. The further along in the overall process one is,

Table 5-1: Elements of Complexity of Boiler Plant Design

PLANT

- Pressurized body
- Components (pumps, fans, etc.)
- Flues and ducts
- Tanks
- Pipes
- Steel constructions
- Concrete
- Electrical System
- Instrumentation
- Heating, ventilation, water, and sewage piping

PLANT DESIGN

Linked chain of
- Process design
- Heat-transfer design and pressurized body design
- Tankage system
- Layout
- Structural
- Electrical
- Instrumentation
- Civil
- Piping
- Complete documentation

COMPUTER-AIDED ENGINEERING

Integrated system of Data-Base Management System(s), hardware platforms, operating systems, applications software, and team

PLANT DOCUMENTATION

- Process flowsheets
- Piping and Instrumentation Diagrams
- Equipment specifications
- Plans, elevations, sections, isometrics
- Random (scrap) views, details
- Bills of material

Adapted by permission from the Center for Integrated Facility Engineering Technical Report 24, Spring 1990.

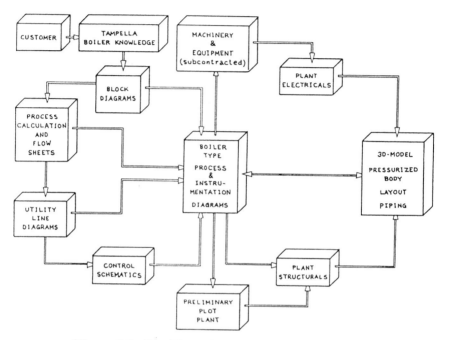

Figure 5-1: The Flow of Data in Boiler Plant Design

Reproduced by permission from the Center for Integrated Facility Engineering Technical Report 24, Spring 1990.

the less the impact on project scope, equipment incorporated, cost, and ultimate delivered performance there will be. It has been estimated that 90% of the cost of a given project is determined during the initial 20% of the product life cycle.

5.3.4. Traditional Approach to Design

In the conventional process, the manufacturing design of the pressurized body begins at the same time as plant design. This results in plant design drawings being oriented to manufacturing as opposed to being made for conveying plant-design information. The manufacturing drawings contain great quantities of detail. It is difficult for the customer who must decide on changes and eventually sign-off on the design to be able to filter out the relevant facts on which decisions can be made. There is a fuzzy border between preliminary and detailed design.

5.3.5. IBPD Approach to Design

IBPD combines in a single knowledge system both the knowledge and hard-won experience that had previously been distributed among many people representing multiple separate disciplines. The resultant design rules permit the interaction among disciplines and immediate (parallel rather than serial) influence of one on the others. The vehicle for visualization was specifically tailored to preliminary design and thus facilitates understanding by the customer. The decision process and getting customer (and other parties, where appropriate) approval(s) are facilitated. The overall approach is to allow the manufacturing-oriented design work to begin after and on the basis of the preliminary-design work rather than the two proceeding in parallel. Thus, the amount of energy and rework is decreased. Time for preliminary design and overall project delivery time are reduced as discussed in the section on benefits below.

5.3.6. Integration with Existing Programs

Like other complex systems, IBPD did not evolve in a vacuum. The numerical process calculations for the boiler running on a VAX had been developed over years and there was no need to convert them to run within IBPD itself. The programs are written in FORTRAN. The output of those programs became input parameters transferred via a Local Area Network (LAN) to the IBPD workstation. In addition, the parametric CAD programs running on the Computervision workstations were well suited to their functions and it was not necessary to replace them. The output of IBPD became the input to those programs via file transfer. These parameter files drive the parametric programs of the Computervision CAD environment. The parametric programs were developed using the CADDS 4X CVMAC programming language. The transfer occurs via a LAN connection to the VAX and then via an RS-232 serial link from the VAX to the Computervision workstation. The idea behind IBPD was to augment the value of rather than to replace existing systems [5].

5.3.7. CAD System and Interface

The version of IBPD used at Tampella includes graphics produced in the S-geometry visualization program of the Symbolics environment. Changes in the geometric representation trigger changes in the underlying symbolic model. When the process moves on to the detail-drawing stage, a file is transferred as input to Computervision. If changes are made in the Computervision representation, they are not automatically reflected back to the symbolic model. The symbolic model has to be updated manually if concurrency is to be maintained.

Tampella decided that even though IBPD could be set up to check for physical interferences, such interference checking would remain only in the Computervision system. One reason for this approach is that the design rules were being written to avoid such interferences in the first place.

In the IBPD case, the process was extended by Tampella to output CNC code based on the detail drawings of Computervision to drive the manufacturing process for metal-tube bending.

5.3.8. Relational Database Interface

Tampella uses the Oracle relational database management system running on Sun machines. The on-line connection between IBPD and Oracle is provided by Design++.

5.3.9. Targeted Capabilities of IBPD

The following are the main capabilities identified as required in the system. Others, and a more detailed description of the overall system are to be found in [4].

- Transfer data among all three design stages: proposal design, preliminary boiler plant design, and detailed design,

- Proposal design and implementation design to be done in separate design groups with defined composition and tasks,

- Rapid formulation of boiler plant knowledge into design rules automating routine design work,

- Shortened design cycles,

- Standardized data flow,

- Enable both customer and boiler plant maker personnel to recognize customer-requested revisions to the original design (for which the customer will be charged),

- Improve the quality of the design,

- Improve the potential for subcontractors to do detail drawings, and

- Facilitate exchange and documentation of plant-design data.

The goal was to get the customer involved at the preliminary design stage including understanding trade-offs, have Tampella get feedback and make revisions, and get sign-off from the customer. This approach obviates the need for the customer to be involved in the detailed-design stage where the maker of the boiler plants is far more proficient than the customer. The needs of the customer, however, have already been satisfied through understanding facilitated by the visualization in the preliminary-design stage.

5.3.10. Development of the Integrated Boiler Plant Design System

During the development process, two successive prototypes were constructed. The library component was called the super knowledge base and the models the work knowledge bases. Figure 5-2 shows a typical IBPD display.

IBPD is used to deal with problems not successfully solved by conventional CAD systems. Examples are layout of equipment in three-dimensional space, maintenance of consistency in data bases, control of large data bases, and managing design revisions. Unlike conventional or parametric CAD, Design++ has the ability to literally reconfigure systems such as a plant, rather than just to resize them.

5.3.11. Upper-Circulation Piping

The upper-circulation pipes and the supporting system for a recovery boiler are shown in Figure 5-3. Components include:

- drum,
- headers,
- intermediate pipes, and
- supporting rods.

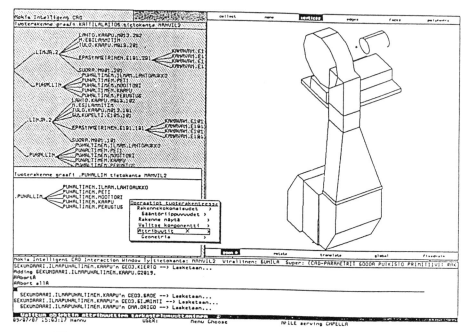

Figure 5-2: IBPD Display of Air Ducts with its Design Windows

Top left is the product structure window and the top right
the 3-D graphical representation. The pop-up menu contains
component-related operations and the window at the bottom
a design-rule trace.

The rods support the pressurized body of the boiler. The performance of the
boiler determines the number, size, and location of all these components. A
body of design rules was created can deal with any Tampella recovery boiler.
Triggering of the design rules may cause more components to be added in which
case more objects are added to the right side of the graph of the hierarchical
product structure and appear in the CAD drawings as well. The product structure is a modification of those of previous projects.

IBPD allows a single part or a small set of components to be visualized if
desired. Figure 5-4 (reproduced from [5]) shows a single circulation pipe as dis-

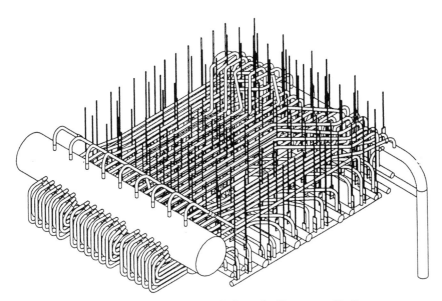

Figure 5-3: Upper Circulation of a Recovery Boiler

Reproduced by permission from the Center for Integrated Facility Engineering Technical Report 24, Spring 1990.

played on the IBPD workstation. Figures 5-5 and 5-6 demonstrate that same pipe as displayed on the Computervision CAD system in (a) a 3-D view, and (b) as the manufacturing drawing, respectively. In the Tampella process, CNC code is generated for a tube-bending machine.

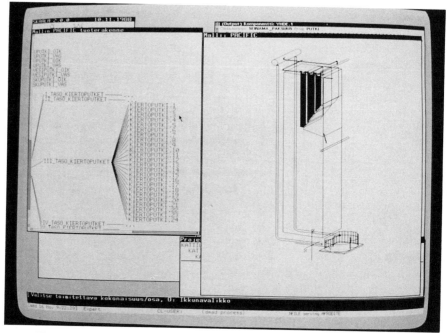

Figure 5-4: IBPD Display of the 3rd Circulation Pipe of III Level 1

Reproduced by permission from the Center for Integrated Facility Engineering Technical Report 24, Spring 1990.

5.3.11.1. Reconfiguration

Maintenance of design consistency is critical. In Figures 5-7 and 5-8, a transformation from having 16 to 28 tubes in the longitudinal section of the secondary superheater is shown. This change was accomplished by changing the *number_of_tubes* attribute through interaction with a dialog box. Once the value of the *number_of_tubes* attribute was changed, all the affected components impacted (as determined through the dependency network) had their attributes recalculated. All of the superheaters, for example, depend on the secondary superheater, and affected attribute values were set to UNKNOWN. With a com-

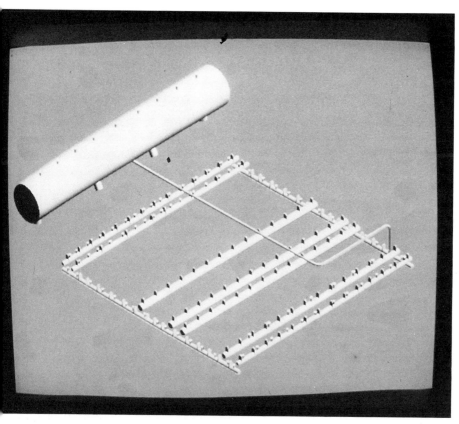

Figure 5-5: Same Selected Circulation Pipe Transferred to and Displayed
in Three-Dimensional Form on the Computervision CAD System

Reproduced by permission from the Center for Integrated Facility Engineering Technical Report 24, Spring 1990.

mand from the IBPD user, the superheaters are then designed again through triggering of the design rules. Note that while some attributes can be locked (therefore not changed to UNKNOWN) and thus not be recalculated), the designer must be careful because under these circumstances it is not possible to guarantee the knowledge-base consistency.

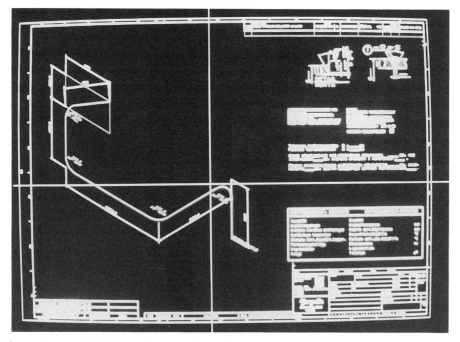

Figure 5-6: Manufacturing Drawing for the Selected Circulation Pipe
Displayed on the Computervision CAD System

Reproduced by permission from the Center for Integrated Facility Engineering Technical Report 24, Spring 1990.

5.3.11.2. Benefits obtained with upper-circulation piping

The decrease in design time obtained using IBPD for the design of the upper-circulation piping was dramatic. Using a completely manual design the process took two months. Using Computervision, including its plant-design software, the design process took two weeks. Using IBPD combined with Computervision CAD, the process only took two days.

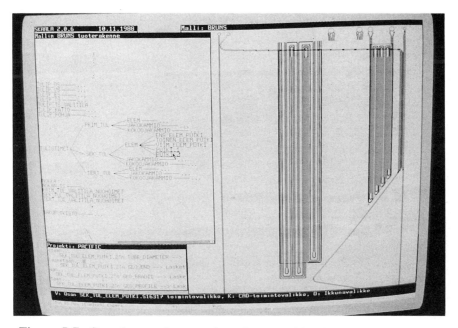

Figure 5-7: Superheaters in case where there are 16 Tubes in the Longitudinal
Section of the Secondary Superheater

5.3.12. IBPD Knowledge

Knowledge with respect to the pressurized section of the boiler (design, manufacture, and erection) is highly codified because of safety considerations. For this reason, this was the first area for which design rules were created in IBPD. Less standardized is the development of the ducts; this is an area of on-going development. In addition to structural design rules, design rules are being developed relating to operational functions such as cleaning of heated surfaces (sootblowing), transport of ash and salts, and mixing of black liquor. The over-

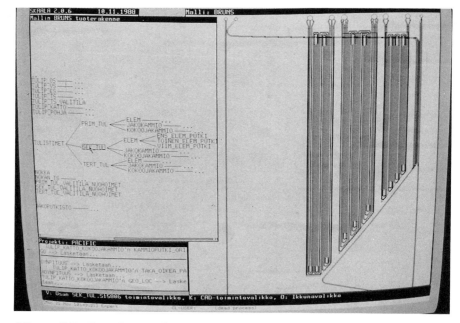

Figure 5-8: Superheaters in case where there are 28 Tubes in the Longitudinal
Section of the Secondary Superheater

Reproduced by permission from the Center for Integrated Facility Engineering Technical Report 24, Spring 1990.

all IBPD now has thousands of plant components with thousands of associated
design rules.

5.3.13. Benefits Obtained on Air-Duct Design

Evaluation of the effectiveness of IBPD was undertaken by comparing the
design of air ducts in a black liquor recovery system using conventional methods
versus using IBPD. A sample air-duct system is shown in Figure 5-9. The

structure is 40 meters high and weighs approximately 70 tons. Air-duct systems for recovery boilers vary so much from plant to plant that using conventional parametric CAD approaches is not possible. Elements of comparison included the work of the project design group leader, the process design, the layout drawings, as well as the detail drawings for manufacture of the air ducts and their installation.

Figures 5-10 and 5-11 show the timing and loading of design effort for conventional (partly using CAD) and IBPD cases. In the latter case the design was new, not having been modified from the conventional case. Not only were fewer design engineer resources utilized in the IBPD case, the design work was finished in a significantly shorter elapsed time. Some comparative figures are presented in Table 5-2.

Other benefits of IBPD are:

- Layout is established, but still flexible,

- Pressure losses can be accurately determined in the preliminary-design phase, thus avoiding the need to oversize fans to allow for a reserve,

- A reliable duct structure is delivered to the customer,

- Only actually required number of service platforms will be included,

- Advantage is taken of integrated duct production, and

- Improved ease of operation is obtained taking into account the connections between ducts and other plant components.

One identified risk was that with IBPD, the high level of duct expertise was incorporated within the system and in a fewer number of engineers than with the conventional approach. The system, however, does aid in helping more junior designers to quickly develop high levels of expertise.

5.3.14. Resources Required

The effort required for IBPD through 1989, including the development of its underlying SCALE tool (now Design++) was fifteen person-years. The steps included were a preliminary design, the first prototype, evaluation, second prototype, trial run, preliminary product version, trial run, and full production. Development work continues intensively in parallel with the production.

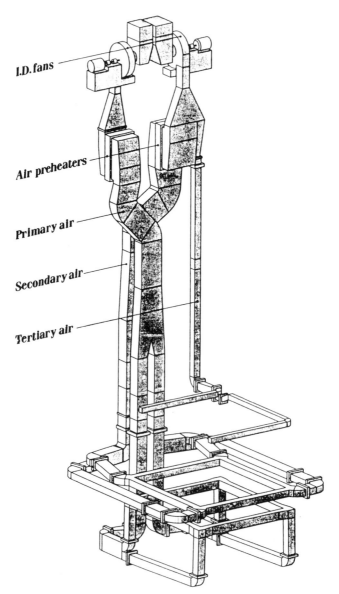

I.D. fans

Air preheaters

Primary air

Secondary air

Tertiary air

Figure 5-9: Recovery Boiler Air Ducting

Reproduced by permission from the Center for Integrated Facility Engineering Technical Report 24, Spring 1990.

Table 5-2: Quantitative Comparison of Conventional Versus IBPD Approaches to the Design of Recovery Boiler Air Ducting

	Conventional	IBPD
Calendar time saved using IBPD	Not Applicable	4.5 months
Design Person Hours	2000	500
Design Leader Person Hours	80	40
Revision of drawings compared to conventional case*	Not Applicable	1 / 5

*Since a drawing is *automatically* produced for each module, the number of original drawings was three times as many for IBPD. The average drawing revision number for IBPD, however, is only 0.1.

Adapted by permission from the Center for Integrated Facility Engineering Technical Report 24, Spring 1990.

5.3.15. IBPD Success Ingredients

Positive factors contributing to the success of IBPD are:

- Strong, active, management support was provided, thus eliminating barriers,
- The management dared to try new methods,
- A small cadre of enthusiastic, motivated engineers was established in the design department in the Tampella Power Industry Division,
- Labor was effectively divided so (a) those in the design department specified the application and the knowledge-systems functionality required to support it, built the application, tested the system, and developed the links to the engineering calculations and CAD, and (b) those at NOKIA, the company providing software services, built the knowledge-system tool,
- Results could be and were measured,

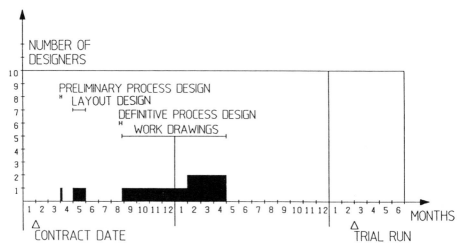

Figure 5-10: Design Effort Loading and Elapsed Time when Conventional
Methods were used to Design Air Ducting

Reproduced by permission from the Center for Integrated Facility Engineering Technical Report 24, Spring 1990.

- Contact was maintained with the primary application users,

- All those working in the design department were trained on the system even if they were not to be users, and

- Competition set up with at least two other Tampella divisions encouraged progress.

These factors are generally applicable in such projects. Critical ingredients are the presence of both user and executive champions.

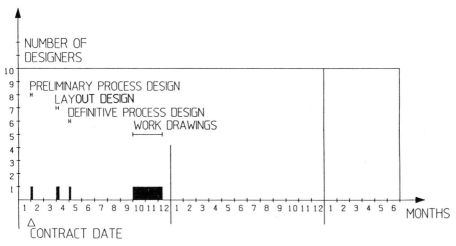

Figure 5-11: Design Effort Loading and Elapsed Time when IBPD
was used to Design Air Ducting

Reproduced by permission from the Center for Integrated Facility Engineering Technical Report 24, Spring 1990.

5.3.16. Summary of IBPD

The IBPD was successfully applied at all three stages of boiler plant design: proposal design, preliminary design, and detailed design. One of the projects in which IBPD was employed was the design of the world's largest soda-recovery boiler. In the case of the ducts, the design effort decreased four-fold from 2000 hours to 500 hours per instance with the elapsed time decreasing 4.5 months. The system is integrated in that design, manufacture, and installation are positively affected and costs decreased. IBPD increases the level of teamwork among the various disciplines. Tampella is pleased that young engineers can be trained as plant designers more rapidly and effectively. The ability to quickly explore multiple alternatives and graphically see their impacts on the overall plant is significantly beneficial. Development continues, both in existing applications and new ones, such as a Piping and Instrumentation Diagram application now underway.

5.4. OVERVIEW OF DESIGN++

The following sections give an overview of Design++, beginning with a description of its open architecture. Later in the chapter, more technical details are given on implementation of the tool. Design++ provides for both the symbolic modeling of the artifact being designed and the glue for the components incorporated in the overall open architecture. The system uses object-oriented programming. The libraries and models built using Design++ are composed of objects. An object may or may not have a physical (geometric) representation, but will have attributes (e.g., length, cost, type of material, etc.) with values and behaviors. The behavior of the object is governed by the attached design rules which in turn may call functions. In a sense, Design++ represents a symbolic spreadsheet for engineering design.

5.4.1. Open Architecture

Design++ uses the functionality of other existing systems rather than replicating such capabilities. As shown in Figure 5-12, Design++ communicates with and incorporates the functionality of CAD, relational data base management systems, engineering-analysis packages, and technical-publishing systems. Thus as these systems evolve, Design++ can automatically take advantage of improvements. Also, functionally equivalent systems can be substituted according to the needs of individual customers (e.g., both Oracle and Sybase are currently supported relational data base management systems).

An important feature is that Design++ was developed to incorporate open architecture from the beginning, rather than this capability being added later.

5.4.2. Layered Architecture within Design++

One gets closer to user solutions by building up layers of functionality. As shown in Figure 5-13, libraries are built upon Design++. These libraries are composed of collections of general domain knowledge upon which collections of customer-proprietary knowledge are layered. Models of artifacts being designed are built on top of the libraries.

IntelliCorp's KEE[TM] [3] is the knowledge-based development tool upon which Design++ is built. Only the frame system of KEE and some interface components are utilized. No use is made of the rule system, assumption-based

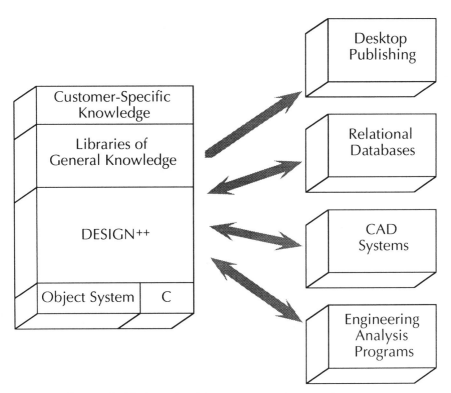

Figure 5-12: Relationships among the Various Components
Linked Together in Design++

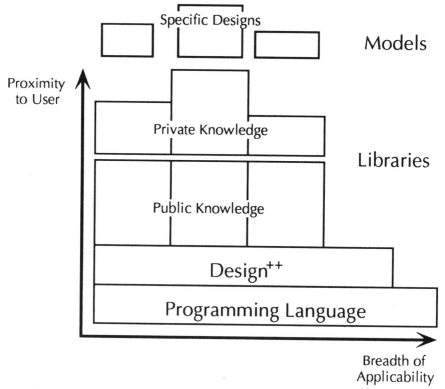

Figure 5-13: Design++ Architecture showing Layering of Knowledge

truth maintenance system or worlds, although Design++ applications could use these additional elements. Design++ runs on both the development version of KEE and Configurable KEE. Some of the features Design++ adds to KEE are:

- Interfaces to CAD systems, relational data base management systems, desktop publishing systems, and external engineering analysis programs,

- Explicit whole-part relationships as used in the model product structure,

- The design rule language with dependency-network control, and

• Geometric reasoning.

5.4.3. Design++ Platform

Design++ and its underlying KEE run on Sun SPARCstations with the SunOS version of UNIX and Sun Common LISP. With the development version of KEE, 24 megabytes of RAM are recommended, while 16 megabytes of RAM are satisfactory for Design++ running on Configurable KEE. A minimal stand-alone system can run with the 207-megabyte hard disk provided on the Sun IPC, but to be practical one needs to plan for over 300 megabytes of hard-disk capacity.

5.4.4. Computer-Aided Design Systems

CAD systems are either tightly or loosely coupled to Design++. A "tightly coupled" system (currently AutoCAD) maintains a two-way link. Thus, not only are the graphics generated by the symbolic model within Design++, but location information about an object moved on the CAD canvas may be transmitted back to Design++ and the underlying symbolic model updated. "Loosely coupled" means that the communication is one way, from Design++ to the CAD system. Currently such connections are available for Computervision and Calma. Other tightly and loosely coupled connections are planned. Once the graphics are "inside the CAD system," the regular CAD functions such as changing point of view, altering layer colors, removing hidden lines, doing shading, etc. are available to the user.

One mode of using Design++ is to use AutoCAD™ as a visualization tool (e.g., for schematics and 3-D layouts) regardless of what system is used for an organization's production CAD system. The production CAD system is then used for producing of the detail drawings, typically based on DXF files transferred from AutoCAD.

5.4.5. Relational Databases

Database access occurs via SQL calls that are incorporated in design rules. Values can also be written out to a relational database from Design++. In addition, behaviors can also be included in the objects in the relational database by storing associated design rules as well. Currently connections to Oracle and Sybase have been created. Other relational databases can be supported as well, as long as the logical and physical paths constituting "the wire" are created. The SQL itself will be the same in most cases, although for efficiency reasons it may be preferable in some cases to rewrite the SQL to tune it to the particular relational database.

5.4.6. External Engineering Analysis Programs

External applications can be accessed over a network; it is not expected that they will necessarily reside on the same machine on which Design++ is running. Examples are process calculations, vibration analysis, discrete or continuous simulation, stress analysis, finite element analysis, and project management.

5.4.7. Description of Design++ and Its Knowledge Representation

As noted above, components of artifacts or systems being designed are objects whose behavior is given by attached design rules. These objects are represented as frames.

5.4.7.1. Objects

Objects have attributes that can have values. An important checking mechanism is that attributes have value classes. Examples are *integer, integer within a given range, real, list, string,* and *(one.of left right)*. If there is an attempt to input a value inappropriate to the value class, the error is flagged and brought to the user's attention. A description of an object - a valve - appears in Figure 5-14. Figure 5-15 shows the characteristics of an attribute including a comment, the default value, the attached design rule, the value class, and the current value. The items from default value on are covered below. The relationship of the displayed object to other objects in the hierarchy is shown in the header information. Figure 5-16 shows a typical window displaying object attributes in a dialog-box format. The determine button is used to trigger an attached design rule to calculate the attribute's value. The lock box is used to lock the attribute and its associated design rule (e.g., in applications using concurrent engineering). Both design rules and locking of attributes are covered below.

Valve

ATTRIBUTE	FACETS	VALUE
Type	**Value.class:** Symbol	**Ball_valve**
Actuator_type	**Value.class:** One.of: Pneumatic Manual Electric	**Pneumatic**
Body_material	**Value.class:** Symbol **Design.rule:** (case (:? My Fluid) (Black-liquor '316SS) (Water 'Bronze))	316SS
Connected_to	**Value.class:** List	Unknown
Cost	**Value.class:** Real	Unknown
Flowrate	**Value.class:** Real	Unknown
Fluid	**Value.class:** Symbol **Design.rule:** (:My-parent Service)	**Black_liquor**
Line_number	**Value.class:** Integer	Unknown
Nominal_size	**Value.class:** Real	Unknown
Pressure	**Value.class:** Real	**500.0**
Tolerance	**Value.class:** Integer	Unknown

Figure 5-14: Frame Representation of the Properties of a Valve

5.4.7.2. Hierarchy of utilization of values

Design[++] is demand-driven in that values of attributes are only calculated as they are required. This approach contributes to the efficiency of systems built using the tool. The order of obtaining values is:

1. if there is a local value use it, otherwise

2. if there is an attached design rule, use it, otherwise

3. if there is an inherited value, use it, otherwise

4. use the local or inherited default value if available, or

5. if there is a value from an external data source such as a relational database or an external engineering-analysis system, use it, or if all else fails: ask the user.

```
[] [index] [^] Component PUMP_BASE
Component PUMP_BASE in Library PLANT
Superclasses:
     BOX in Library GEOMETRIES
     PUMP_PART in Library PLANT
Subclasses:

Comment: <<not specified>>

GEO_LOC
  Comment: ((:ENGLISH "Object location in cartesian co
ordinate system") (:FINNISH "Objektin sijainti suorakul
maisessa koordinaatistossa"))
  Default: (0.0 0.0 0.0)
  Design Rule:
(! SELF GEO_LOC
   (COND
      ((= 1 (:? PUMP ^INDEX))
       (:LEFT-OF (:?1 LEFT_COLUMN_1)
                 GEO_LOC
                 (+ 2000.0 (:? MY GEO_WIDTH)))
       (:FRONT-OF (:?1 LEFT_COLUMN_1)
                  GEO_LOC
                  (+ 3500.0 (:? MY GEO_HEIGHT)))
       (:ABOVE (:?1 LEFT_COLUMN_1) GEO_LOC))
      ((= 2 (:? PUMP ^INDEX)) (:RIGHT-OF (:?1 RIGHT_COLUM
N_1)
                               GEO_LOC
                               2000.0)
                 (:BEHIND (:?1 RIGHT_COLUMN_
1)
                          GEO_LOC
                          500.0)
                 (:ABOVE (:?1 LEFT_COLUMN_1)
                         GEO_LOC))))
  Valueclass: (LIST.OF FLOAT)
  Value: (0.0 0.0 0.0)
```

Figure 5-15: A Full Display of an Attribute of a Typical Component
Showing Various Facets

Note that access to relational databases occurs through design rules.
Another way to view the integration of the various system components is to
look at the representation of objects. Figure 5-17 illustrates the data,
knowledge, and graphic levels.

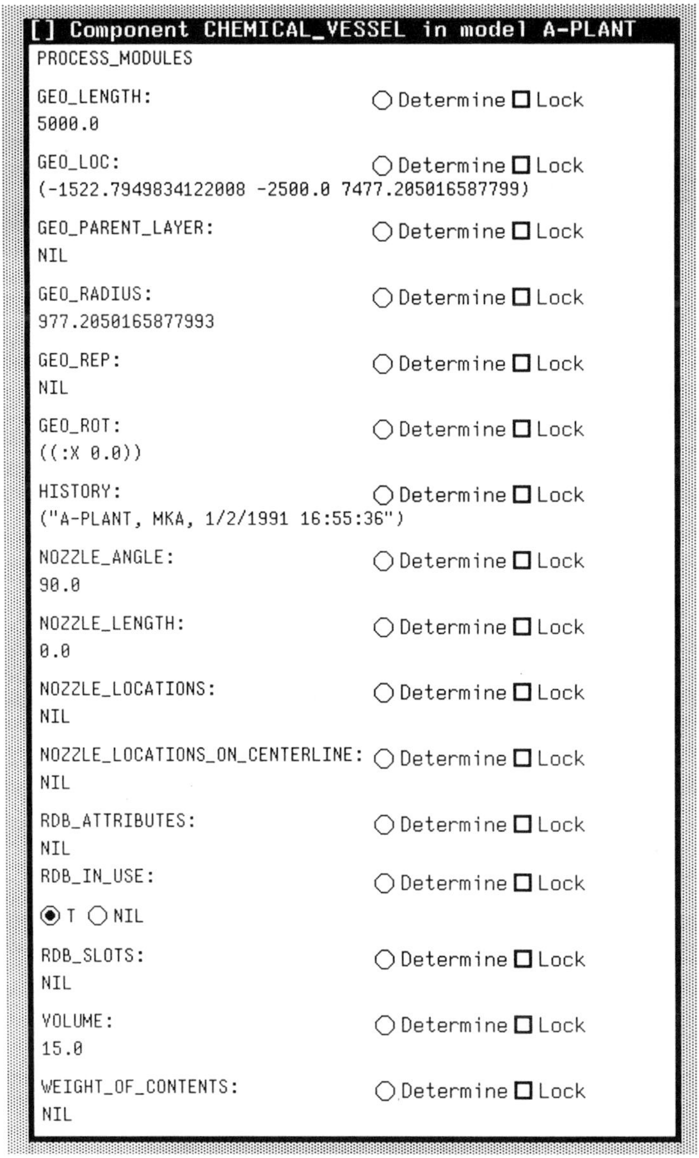

```
[] Component CHEMICAL_VESSEL in model A-PLANT
PROCESS_MODULES

GEO_LENGTH:                    ◯ Determine ☐ Lock
5000.0

GEO_LOC:                       ◯ Determine ☐ Lock
(-1522.7949834122008 -2500.0 7477.205016587799)

GEO_PARENT_LAYER:              ◯ Determine ☐ Lock
NIL

GEO_RADIUS:                    ◯ Determine ☐ Lock
977.2050165877993

GEO_REP:                       ◯ Determine ☐ Lock
NIL

GEO_ROT:                       ◯ Determine ☐ Lock
((:X 0.0))

HISTORY:                       ◯ Determine ☐ Lock
("A-PLANT, MKA, 1/2/1991 16:55:36")

NOZZLE_ANGLE:                  ◯ Determine ☐ Lock
90.0

NOZZLE_LENGTH:                 ◯ Determine ☐ Lock
0.0

NOZZLE_LOCATIONS:              ◯ Determine ☐ Lock
NIL

NOZZLE_LOCATIONS_ON_CENTERLINE: ◯ Determine ☐ Lock
NIL

RDB_ATTRIBUTES:                ◯ Determine ☐ Lock
NIL
RDB_IN_USE:                    ◯ Determine ☐ Lock
◉ T ◯ NIL

RDB_SLOTS:                     ◯ Determine ☐ Lock
NIL

VOLUME:                        ◯ Determine ☐ Lock
15.0

WEIGHT_OF_CONTENTS:            ◯ Determine ☐ Lock
NIL
```

Figure 5-16: Window Representation of Component's Attributes Dialog Box

Levels of Knowledge

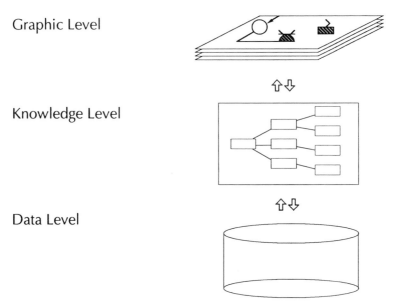

Graphic Level

Knowledge Level

Data Level

Figure 5-17: Systems Integration of Object Representations

5.4.7.3. Projects, libraries, and models

Designs are organized by project. Projects consist of libraries and models. The Design++ library has three root nodes: the index, the assembly, and the part. The index contains basic descriptive and historical information about the library. The assemblies and the parts both form the basis of classification hierarchies for defining components of a product. Assemblies provide an organizing function, both for decomposing a product structure and for performing assembly-related processes such as calculating geometry. Assemblies do not have geometry inherently and are only "displayed" in terms of their component parts which do have their own inherent geometric representations. Attributes can be added at any level of the tree in either the assembly or part hierarchy and passed down to objects "below" them. An example of a window display of a library is shown in Figure 5-18 for a simple process plant. A given application can have multiple libraries associated with it. A model is a product structure representing the specific artifact being designed. The root of the whole/part product-structure tree is an assembly that appears in the assembly hierarchy of an appropriate library.

Components in the library know how to size, connect, or perform other functions on themselves and related components. A steel beam, for example, can know how to size itself based on characteristics of its support, its length, the load it will bear, contracting-organizations specifications, building codes, and other factors. With respect to connectivity, a pump can know to what downstream components (say a valve, a pipe, and a vessel) it is connected and behave accordingly. Components contained in relational databases (or flat files) can be represented by template parts in the Design++ libraries so they will have intelligence as well. The system can also be set up so a design rule associated with a given object/attribute in an external data base can be stored with that object/attribute. The objective is to provide knowledge in a reusable form.

5.4.7.4. Design rules

The design-rule system provides tools for expressing design knowledge and plays a central role in maintaining knowledge consistency. Design rules in Design++ are deterministic engineering expressions rather than the "if-then" production rules of traditional expert systems. The maintenance of large knowledge bases is markedly simplified by this approach as compared to traditional rule-based systems. Design++ design rules are inherently transparent in their representation of knowledge and provide an automatic explanation system. A design rule is attached to an attribute by compiling it in a rule-editor buffer. A typical rule might read:

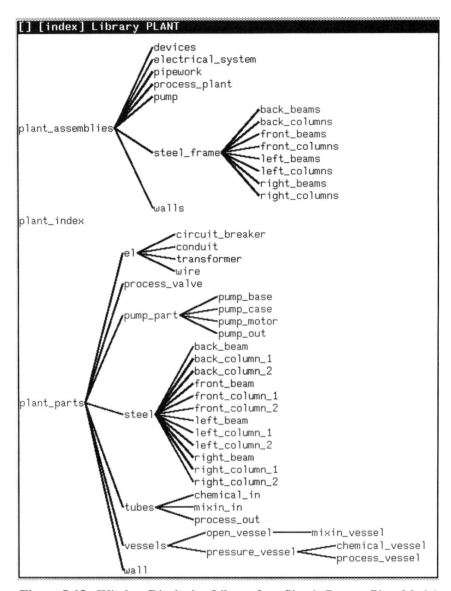

Figure 5-18: Window Displaying Library for a Simple Process Plant Model

```
(:! unit_back_left_corner_column geo_loc
     (:right-of main_back_left_corner_column geo_loc
      5500.0))
```

which means that the geometric_location *(geo_loc)* of the given units *back_left_corner_column* is to the right of the *main_back_left_corner_column* by 5.5 meters (5500 millimeters). The ":!" at the beginning of the design rule indicates to Design[++] that the expression is a design rule. The syntax of design rules is that of LISP and any legal LISP expression can be included. The process is simplified by having a series of LISP macros, for example those in the geometric domain, such as :above, :right-of, :left-of, :behind, etc. These macros are abstractions that allow a designer to express relationships in a very natural way.

Geometric relationships are not the only ones that can be expressed in design rules. For example, one might have:

```
(:! transformer input_current
     (:? (:? own input_component)
      output_current))
```

which means that the *input_current* of the *transformer* is equal to the *output_current* of the *transformer's* own *input_component*. This may seem obvious, but the system needs an explicit representation of this knowledge. There is no way for a conventional CAD system to express such a relationship. Design rules include reference clauses to facilitate designating objects and attributes to be related. For example,

```
(:? transformer current_capacity)
```

when used within a design rule would deliver the value for the *current_capacity* attribute of the *transformer* component.

Just because design rules are engineering expressions rather than "if- then" production rules of traditional expert systems does not mean that conditional expressions are not permitted. For example,

```
(:! door installation_time
     (if (< (:? own height) 12)
      3
      4))
```

means that the *installation_time* for a door in the given model is three units of time if the *door's* own height is less than 12 and four units of time otherwise.

A useful strategy can be the application of "generic design rules" which can be passed down to wherever they are needed and then instantiated according to the values of one or more local attributes. For example, a generic design rule may cause a given component to be added, substituted, or deleted according to the value in an "operation-to-be-performed" attribute.

Design rules can operate with the rule trace on or off. This toggle is accessed through the Design++ utilities menu (at the Design++ icon bar). With the trace on, the behavior of the system is shown in terms of the processing going on, but naturally the system will operate more slowly.

Design rules, including the feature of direct- and indirect-reference clauses, constitute an extremely powerful capability of Design++ in facilitating object-oriented programming as opposed to doing it in object-oriented LISP alone.

5.4.7.5. Models and their product structures

Models are product structures for artifacts being designed. An example is shown in Figure 5-19. Nodes in the product structure do not inherit from higher ones. For example in the *electrical_system* assembly component of the simple plant model shown in Figure 5-19, the part components such as *transformers* and *circuit_breakers* do not inherit any of their attributes from the *electrical_system*. This does not prevent one from having attributes of similar or the same name for a given purpose. For example, if one wants to aggregate costs of the assembly, one could have a design rule:

```
(:! assembly combined_cost
        (+      (:? part_1 cost)
                (:? part_2 cost)
                (:? part_3 cost)))
```

where whether the assembly's *cost* attribute were *combined_cost* or *cost* would make no difference. This is an example only, normally one would loop through (:parts assembly).

5.4.8. Interaction with Relational Databases

SQL queries can be included in design rules. In the following design rule, the diameter of a connection tube of a pre-heating section of a boiler furnace is calculated by:

```
(:! connection_tube diameter
        (* 2 (SQRT (/ (:? boiler_plant steam_capacity)
                      (/ (:? my flow_speed)
                         (rdb-sql "Select Distinct Value from
                      Physical_factors where
                      Key = Water_density" 'STD))))))
```

The *steam_capacity* value is obtained from the object *boiler_plant*. The *water_density* value is retrieved from the table *physical_factors* in a relational

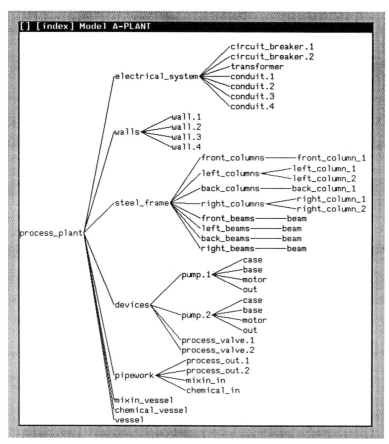

Figure 5-19: Product Structure for a Simple Three-Vessel Plant Model

data base. Note that if there is change in the *steam_capacity* value later on in the process, the system automatically maintains consistency through the dependency network inherent in the model and the *diameter* of the *connection_tube* is automatically redetermined.

Libraries of components can be stored in relational databases. Data according to ASME, DIN, and other standards can be made available in this form as can data from manufacturers' catalogs (e.g., for pumps, valves, instruments, etc.).

5.4.9. User Interface

Besides the tree displays of libraries and product structures, the object/attribute/value displays (full, dialog-box, or other), and the rule-trace (input/output) window, one can create dialog boxes, pmenus (including those with icons representing selection types), and active images (using KEEpictures). Active images can be vertical or horizontal bar graphs, traffic lights, digital meters, thermometers, dials, etc. In addition to display, some of the KEEpictures are actuators and can provide input to the values of attributes as well. Sets of images can be grouped together by attaching them to image panels.

5.4.10. Reports as Models

One of the most powerful features of Design^{++} is that reports are models with their own product structures. As with product structures for artifacts being designed, those for the report are constructed components in an appropriate library. The objects are concerned with characteristics of text including the formatting thereof, the intermixing of graphics and text, and non- geometric attributes. Among the components of a report can be textual descriptions, specification tables, bills of material, proposals, purchase orders, isometric piping diagrams with associated takeoffs, and even tailored installation or operating instructions. The design rules will deal with different attributes such as cost, weight, and overall component quantities. A simple product structure for a sample report is shown in Figure 5-20.

One creates a report model by clicking on the model canvas as one does to initiate any model. Then either a report default product structure is read in from a file or one is created interactively. The model of what is being designed is then attached as a secondary model. Whenever the secondary model is changed and a new report is generated, the report will reflect the new conditions. This dynamic linking of the report to the model allows concurrency to be maintained and simplifies document management. One application that can be accomplished within the context of a report is data translation. An example is translation of one format to another of input to different versions of software with the same function (e.g., input to alternative engineering analysis programs).

Outputs of the report component of Design^{++} can either be in ASCII files or desktop publishing programs such as FrameMaker or Interleaf. The latter will support the intermixing of text and graphics. The capability of automatically producing an updated version of a proposal, set of specifications, or other document is powerful indeed. A partial sample page of report output is shown in Figure 5-21.

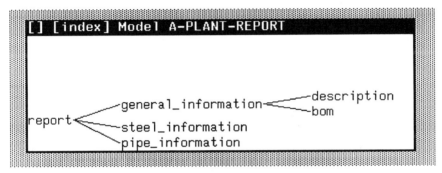

Figure 5-20: Product Structure for a Report

5.4.11. Design-Rule Editor

To facilitate the input of design rules, an editor was created as shown in Figure 5-22. The developer is given selections among the various primitives available (e.g., mathematical or LISP functions), a list of all the objects in the model, and for the selected object, a list of the available attributes. Thus a design rule can be constructed using a "point-and-click" procedure. The text window displayed is a LISP editing buffer. The "star-like" figure below the top set of menu boxes allows movement via point-and-click around in the buffer. Regular Emacs can be used with all of its functions as well. To get into the rule editor, one clicks on an object in the library or product-structure trees and gets a menu which has rules as one choice on it (others being display, delete, rename, etc.). If there are one or more existing design rules for the object then a menu is displayed of those rules. One picks the desired rule, and it appears in the design-rule editor window. Rules are attached to the given attribute when the selected rule is compiled. The rule-editor menu on the right side just above the text window has functions such as compiling the rule, saving the buffer, and exiting the design rule Editor. The rest of the design rules in that file are accessible in the given buffer as well.

REPORT OF A-PLANT

GENERAL INFORMATION

The chemical and mixin vessels are located on the same floor if there is enough room for them. If the maximum width of the steel structure were exceeded a new floor would be created automatically and all the vessels be located on the one-for-each-floor basis.

The pumps, pipework and walls are located around the vessels and steel structure. Different design alternatives can easily be tested because the model reacts dynamically.

DESCRIPTION

The plant consists of vessels, pipework, pumps and steel structure.

Some dimensions of the process vessel:
- outer diameter 2523 mm
- length 5000 mm
- volume 25.0 m**3

BILL OF MATERIALS

PROCESS_PLANT
    ELECTRICAL_SYSTEM
        CIRCUIT_BREAKER.1
        CIRCUIT_BREAKER.2
        TRANSFORMER.1
        CONDUIT.1
        CONDUIT.2
        CONDUIT.3
        CONDUIT.4
        TRANSFORMER.2
    WALLS
        WALL.1
        .
        .
        .

STEEL INFORMATION

Total price: 317 FIM

The steel structure consists of the following items:
    FRONT_COLUMN_1.1   HEM240   6000.0 mm
    FRONT_COLUMN_1.2   HEM240   6000.0 mm
    LEFT_COLUMN_1.1   HEM240   6000.0 mm
        .
        .
    BACK_BEAM.2   IPE180   3000.0 mm
    RIGHT_BEAM.1   IPE180   6000.0 mm
    RIGHT_BEAM.2   IPE180   6000.0 mm

Total length: 96000 mm

The steel structure consists of the following specifications:
    HEM240   60000.0 mm
    IPE180   36000.0 mm

PIPE INFORMATION

The pipework consists of the following items:
    PROCESS_OUT.1   diameter 65.0 mm   10717.3 mm
    PROCESS_OUT.2   diameter 65.0 mm   6317.3 mm
    MIXIN_IN   diameter 40.0 mm   19949.4 mm
    CHEMICAL_IN   diameter 50.0 mm   18376.8 mm

Total length: 55360 mm

The pipework consists of the following diameter groups:
    diameter 40.0 mm   19949.4 mm
    diameter 50.0 mm   18376.8 mm
    diameter 65.0 mm   17034.6 mm

Figure 5-21: Sample Page of Report

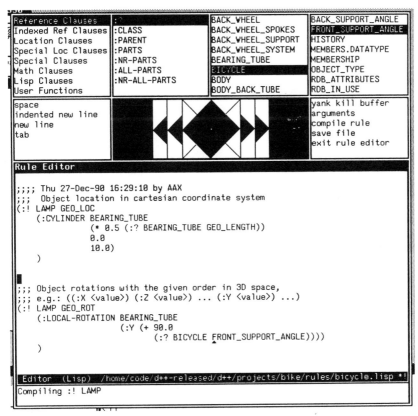

Figure 5-22: Sample Dump of a Design-Rule Window

5.4.12. Constructing Component Attributes from Knowledge/ Data Sources Using the Dependency Network

Design[++] can deal with very complex systems. Values for attributes can come from within one of the Design[++] knowledge bases, external engineering analysis systems, relational databases, or appropriately formatted flat files. Figure 5-23 shows the dependency network displaying relationships among the attributes of a steel beam. If the value for the depth of the beam is requested, the system will traverse the network as indicated by the arrows. As reflected in the network, if the beam's material's modulus of elasticity is changed, its moment of inertia, its section, and its depth will be impacted, but the beam load point, length, and allowable deflection will not. A similar example is discussed more completely in [2].

5.4.13. Reconfiguration Capabilities

As an example of the power of Design[++], Figures 5-24 and 5-25 show a simple, three-vessel, process plant before and after an attribute of one of the vessels was changed. The plant includes a 25 cubic-meter process vessel on the floor of the plant. In Figure 5-24, the two other vessels are each 15 cubic meters in volume. In Figure 5-25 the volume of one of the upper vessels has been changed to 20 cubic meters. Since there was a width constraint incorporated into one of the design rules, the two upper vessels can no longer rest side by side, but an additional floor must be added. This does not just change the shape of the plant building and the form of the structural steel supporting the upper vessels; it affects the pumps and electrical system as well. The pump in the foreground must now output enough pressure to raise the fluid to a higher elevation. It, in turn, requires more power as reflected in a larger circuit breaker/motor starter and now two transformers (in the upper-rear corner) instead of one. Such reconfigurations cannot be accomplished (except in an individual case by case way) with parametric design systems.

5.4.14. Concurrent Engineering

Preliminary design and all the steps that follow benefit from concurrent engineering. The earlier in the overall process that a required or desirable change in design is made, the less the negative impact on both cost and delivery schedule. Concurrent engineering is of two flavors, multidisciplinary and life cycle. Multidisciplinary refers to the engineering domains involved in the design of the product itself. For example,

- Structural,
- Mechanical,
- Electrical,

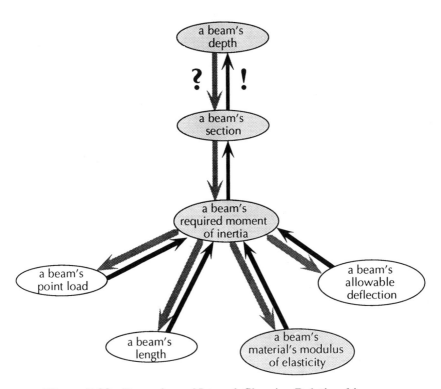

Figure 5-23: Dependency Network Showing Relationships among
Steel-Beam Attributes

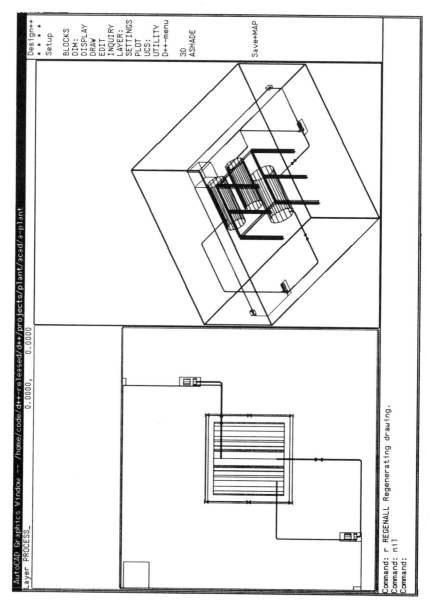

Figure 5-24: Three-Vessel Process Plant Prior to Changing an Upper-Vessel Volume

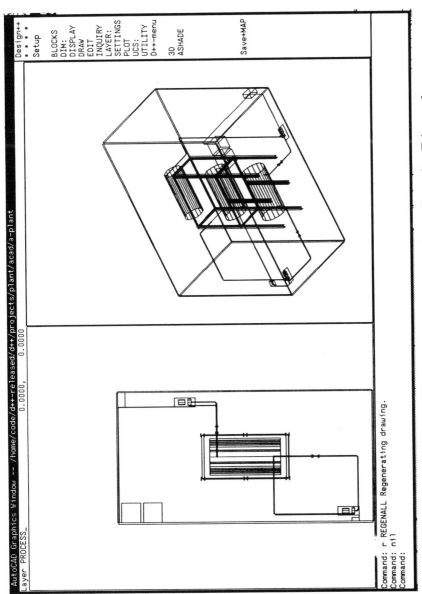

Figure 5-25: Plant shown in Figure 5-24 after Reconfiguration Triggered by Change in an Upper-Vessel Volume

- Instrumentation, and

- Safety and other compliance

all have their interdependencies recognized and satisfied within the overall context of the product. As noted, building code and other compliance regulations fit in this category. Using Design++, each discipline's requirements are considered in parallel rather than serially.

Life-cycle concurrent engineering is a longitudinal view. Rather than creating a product design and "throwing it over the transom," life-cycle requirements are considered in a "Design for ___ " approach. Thus, constraints of

- manufacturability (e.g., materials impacting tooling),

- transportability (e.g., a plant module cannot exceed these given dimensions and weight),

- testability (e.g., testing ports must be included), and

- maintainability (e.g., service-access requirements such as a specified value should not exceed a given height)

can be respected and have their requirements incorporated as surrogate representatives of their associated engineers.

Another capability that can be included is that of design knowledge capture. This can include the documentation of such considerations as why certain decisions (e.g., selection of material) were made, what was the justification for not making alternative choices (e.g., we would have chosen material x if it had not cost twice what the selected material cost), and what was the design intent (e.g., we selected a given thickness of concrete pad between building floors not just for strength, but for aesthetic reasons).

5.4.14.1. Implementation of concurrent engineering

A Design++ product-structure model can contain design rules incorporating any and all engineering relationships to be maintained. The overall model is called the Control Model. Individual disciplines or life-cycle groups "check out" relevant sections of the model in the sense that they can change the design rules and attribute characteristics for which they are responsible. They are actually working with the entire model, a copy called the Work Model, but are able to change only their own sections of it. All other attributes are or may be locked. Locking of attributes (which also locks the attached design rule, if any) is one way to ensure that a given attribute or design rule is left unchanged. Thus certain aspects of the electrical system design would remain inviolate when changes in design are being made by a member of the mechanical team.

The segmentation may not be strictly along given component lines, but may cross those boundaries. For a given component, a set of functional attributes may "belong" to one group (e.g., the cost-estimation engineers) while the geometric attributes are the responsibility of another. If there are attributes or design rules which require input from multiple areas, then negotiation occurs.

The control model is the responsibility of the program manager for the overall design. When periodic integration occurs, he or she arbitrates any conflicts. Version control is applied to ensure that the status of given models is known and the most current model is clearly identified.

5.5. TECHNICAL BASIS OF DESIGN++

To provide more insight into how Design++ was implemented, additional information is provided about its technical basis in the following sections.

5.5.1. Design-Rule Implementation

Design rules are written in the design-rule language which is a macro language built on top of LISP. In addition to geometric macros, there are both direct and indirect referencing macros relating to the attributes of either the same or other components (objects). Since the design-rule language is embedded in LISP, the power of LISP is available transparently if needed. Because the relationships most users need are provided by the supplied macros, arithmetic expressions, and simple conditionals, extensive use of the richness of LISP itself is not usually required. As noted above, the intelligent design-rule editor further lowers the effort required from the end- user to write design rules.

Control within the design-rule system is demand-driven. The value of each attribute is not determined until there is a need for it. This approach results in the minimum necessary computation thus improving system performance. The design-rule language is declarative and order independent. The designer can concentrate on defining the relationships among components and their attributes without worrying about (a) the flow of control as in traditional programming environments, or (b) the order that rules were added in rule-based environments.

Demand-driven control is implemented using daemons monitoring the values of attributes. A daemon is attached automatically to an attribute along with its design rule assignment. The value-monitoring daemon is activated whenever an attribute is referenced. This is implemented using the get.value facility of KEE. If the (local) value of the attribute is UNKNOWN, the monitor determines the value by invoking the design rule (if there is one) attached to the attribute. Actually the daemon calls the LISP function into which the design rule has ultimately been compiled. If there is no design rule to be triggered, the value is sought from an alternative source as covered previously.

5.5.2. Locking Attributes

The design-rule system also offers a mechanism for locking the values of attributes. Locking disables the value changes from both the user and consistency-retaining removal propagation. Attributes can only be locked if a value is in place. When an attribute is unlocked, the attached design rule (if there is one) is invoked and a new value is calculated. This is done to allow the overall design to have maximum consistency. Parenthetically, if a new design rule is compiled for the given attribute, the user is warned that the value of the attribute needs to be unlocked before its design rule can be changed. Locking is used for freezing approved design phases. It also supports concurrent engineering by preventing designers from changing values of attributes outside their own discipline.

5.5.3. Maintaining Consistency

Relationships between attributes are defined by references to other attributes through design rules. During rule execution, referenced attributes are dynamically connected with bidirectional arcs to form a dependency graph (network). This dependency network is not a generalized constraint system. The dependency network is used later for maintaining the consistency between attributes of components. A change in the value of an attribute causes a dependency-directed removal propagation based on the dependency graph. During the removal propagation, values of all directly and indirectly dependent attributes are removed. Propagation of value changes is implemented with another value-monitoring demon which is attached automatically when dependencies are formed. To improve the performance, as noted above, the redetermination of values of dependent attributes is demand driven; values are not determined until they are referenced.

The consistency between attributes is not only maintained during value changes but also during design-rule modifications and certain product-structure operations. Such consistency-critical operations, for example, are the creation, copying, and deletion of a component.

5.5.4. Multiple Inheritance

Attributes are added to an object if they are needed. For example, only those objects for which a *geo_type* has been specified inherit other geometric properties. In the case of a box, for example, the attributes *geo_loc* (geometric location such as (1.0 90.0 2.0)), *geo_length*, *geo_width*, and *geo_height* are added.

5.5.5. User Functions

A number of user functions and macros have been defined to facilitate the design rule language. A few examples are:

- prompt-and-read-expression (prompts for an input with an optional output message)
- make-pop-up-menu
- make-pop-up-menu-until-choice-is-made
- vec-plus (calculates the sum of two vectors)
- mat-mult (calculates the product of two matrices)
- model-output (generates an ASCII file from an active model in component attribute value format)
- length-of-centerline (returns the length of a list of corner-point values)
- rdb-sql (executes an SQL clause in a relational data base and returns the results using an on-line link between Design++ and a relational data base)
- bom (pretty prints a product structure to be used for such purposes as printing a bill of materials)
- area (calculates the area bounded by a point list)

- from-menu (displays a menu and returns a single menu selection from the user)

- from-multiple-menu (displays a menu and returns multiple menu selections from the user).

Additions are constantly being made as new applications are developed or existing ones expanded or refined.

5.5.6. Interfaces to CAD Systems

As noted above, CAD systems can be either tightly or loosely coupled, with AutoCAD being the only tightly coupled system currently. The latest AutoCAD for the Sun is Version 10 which does not allow for inter-process communication so transfer in both directions is via file. This restriction to file-transfer mode will be removed with the release of AutoCAD Version 11 for the Sun in mid-1991. Design++ generates an AutoLISP file which is input to AutoCAD.

5.5.7. Interfaces to Relational Databases

There are two components to making a connection to a relational database. One is the generation of the SQL itself, and the second is the connection to the actual database. The SQL generated in the design rules within Design++ is suitable for any relational database using standard SQL. Connections have thus far been produced for Oracle and Sybase, with others developed as needed.

5.5.8. Interfaces to External Engineering-Analysis Applications

The foreign-function-call facility of Common LISP is used to provide interfaces to external engineering analysis applications. The formats, of course, depend on the input/output requirements of the external program. There is no requirement that such engineering-analysis programs run on the same workstation as Design++; they can be executing on a different node in the network.

5.5.9. Interfaces to Technical Publishing Systems

Interfaces to desktop publishing systems such as Interleaf or FrameMaker are handled through file transfers. They can include the importing of arbitrary bit maps from the workstation screen or plot files from CAD systems, so text and graphics can be intermixed.

5.6. ADDITIONAL MODEL-BASED REASONING APPROACHES

The reasoning based on the product structure with attributes represented in the frame system is itself model-based reasoning. One can also have reasoning related to the design process, perhaps to evaluate the effectiveness of the design. For example, based on the same product structure, one can evaluate whether the design's "buildability" is realistic by looking at the resultant schedule (or alternative schedules) using a project-management module. One can also do simulation. As an example, one of the organizations interested in the potential of Design^{++} for intelligent network configuration and management suggested that it would be helpful to see an Ethernet simulation based on the tool. While in practice, one would connect Design^{++} to a simulator rather than do simulation within the Design^{++} application, the fact that one can do such model-based reasoning with Design^{++} underscores the power of the approach. Figure 5-26 shows a trace of design rules during a simulation run.

By monitoring feedback from an operating network, one can also provide intelligent network management. This could include the ability to reconfigure a network on the fly to meet current conditions (e.g., route messages through alternative paths if a link is broken or overloaded, bring back-up components and systems on line, etc.). Such a system could also make recommendations for how to augment or change the network as trends were detected in ongoing traffic.

Design^{++} can be applied to planning and replanning in the sense of dynamically changing the design. Now not only "products" can be designed, "processes" can be designed as well. This particularly fits in with the concurrent-engineering approach inherent in Design^{++}.

```
PC's SCRIPT_PROCESSOR --> Using rule...
Number of frames in message is: 50
Time increment is: 5
    PC's ADD_AGENDA_EVENT --> Using rule...
Have arrived in add_agenda_event for #[Unit: PC.S14 E3]
    ETHERNET_SAMPLE's SORT_AGENDA_LIST --> Using rule...
SCRIPT_PROCESSOR
Current clock_tick before case is 1
Current check_clock before case is 1
    MAIN_FRAME's SCRIPT_PROCESSOR --> Using rule...
Number of frames in message is: 50
Time increment is: 5
    MAIN_FRAME's ADD_AGENDA_EVENT --> Using rule...
Have arrived in add_agenda_event for #[Unit: MAIN_FRAME.S16 E3]
    ETHERNET_SAMPLE's SORT_AGENDA_LIST --> Using rule...
SCRIPT_PROCESSOR
Current clock_tick before case is 1
Current check_clock before case is 1
    WORKSTATION's SCRIPT_PROCESSOR --> Using rule...
Number of frames in message is: 50
Time increment is: 5
    WORKSTATION's ADD_AGENDA_EVENT --> Using rule...
Have arrived in add_agenda_event for #[Unit: WORKSTATION.S15 E3
```

Figure 5-26: Design-rule Trace for an Ethernet Simulation

5.7. TRAINING CONSIDERATIONS

Design++ facilitates training in two ways. First, much of the design expertise needed for using an application and associated knowledge libraries built with Design++ is already incorporated. For example, smart components know how to size themselves and geometric as well as functional relationships are automatically maintained through dependencies. The dependencies are incorporated in design rules that are engineering expressions. Second, the Design++ symbolic model is easy to browse through for training purposes. The design rules are incorporated as attributes of their related objects and facilitate understanding of what is going on (e.g., why a component is of a given size or why there are a given number of components). Such a glass-box approach facilitates getting up to speed quickly and shortens the time to productivity.

The initial training consists of a one-week structured program that includes the opportunity for prototyping on a candidate problem for each of the involved organizations. One excellent mechanism is that of the Apprenticeship [1] in which development occurs with half or more of the activity being done as a joint activity between Design Power, Inc. and the customer/client. One important underlying theme is the transfer of technology from Design Power Inc. to the customer.

The experience at Tampella shows that on-the-job training works. Users are engineers in disciplines directly related to boiler plants (as opposed to having been trained in computer science or its artificial intelligence branch). They are productive in using IBPD within one to two weeks of initiating work with the system. One of the objectives for IBPD is using it as a training tool.

5.8. PROBLEM SELECTION

The maintenance of confidence necessary for project continuation requires that an early quick win be achieved. Problems must be selected so they are initially simple, concrete benefits can be measured, and early successes can be built upon incrementally. The importance of providing the ability to continue production with ongoing payback in parallel with making incremental improvements and additions such as was done at Tampella cannot be overstressed. The principle of setting up a quantitative benefits realization program where the management responsible for a given area are judged by their success in achieving measurable benefits is key.

5.9. KNOWLEDGE PUBLISHING

As applications requiring libraries of components develop over time, knowledge publishing will come into play. Libraries of components such as standard beams that know how to size themselves, components such as fasteners from manufacturers catalogs in machine-readable forms, and compliance regulations will be made available. Any appropriate standards can be included (e.g., AISC, DIN, ASME, UBC, etc.). Design Power Inc. has been exploring this area including such information sources as the Construction Criteria Base available on CD-ROM. This is part of the strategy of layering knowledge in that more and more knowledge will be available. General design knowledge might come from technical books or papers as well as from specifications provided by manufacturers. Then, at the top level, proprietary knowledge representing a given organization's way of doing things and competitive edge is incorporated. Such knowledge could include process knowledge, esthetic approaches, manufacturing methods, design expertise, types of manufacturing facilities available, proprietary components available, lead times for delivery of components, or anything else not shared by a given organization with others.

5.10. FUTURE GROWTH PATHS

Among the areas currently underway are (a) the move of Design^{++} to operate on KEE 4.0 which includes the ability to work in X Windows and provide for color in the Design^{++} canvas as well as provide for more efficient storage of objects within knowledge bases, (b) improvements in the user interface, and (c) the integration of additional CAD systems in a tightly coupled mode. Other areas being considered are (a) movement of more of the user interface for the application onto the CAD window as opposed to the Design^{++} window, (b) explorations in the use of a client-server architecture where Design^{++} would execute on a Design^{++} server running on a workstation and users would be running individual CAD programs on personal computers that would communicate with Design^{++} over a network, and (c) the movement of Design^{++} to operate on a C-based as opposed to a LISP-based object system.

5.11. CONCLUSIONS

Knowledge-based approaches are adding significantly to CAD-system abilities by putting meaning behind CAD graphics. CAD systems are mainly utilized currently as drafting systems rather than incorporating design knowledge. In fact, the majority of CAD systems are used by drafting staff as opposed to engineers. Design^{++} does not just add knowledge and expertise from one engineering discipline; all the disciplines working on a project can be represented. This concurrent engineering approach also includes life-cycle considerations such as Design for Manufacturability and Design for Maintainability.

The open architecture of Design^{++} where the functionalities of CAD, desktop publishing, relational databases, and engineering-analysis programs are provided by interfacing (usually standard) modules provided by other vendors is powerful. Thus, functionality that can be effectively provided (and further improved) by others is not replicated.

The design-rule language with its referencing capabilities and incorporation of macros and functions to facilitate the adding of engineering knowledge to a given application is powerful indeed. The demand-driven dependency network maintains consistency in the model. The layered approach within Design++ incorporates both generally available and proprietary knowledge.

Design++ facilitates engineering productivity because designers can make selections from libraries of assemblies and smart parts that know how to design themselves. The easily understood product structures can quickly be changed to meet new demands for new products or expansions of a given product. A significant approach is that reports have models with design rules and product structures as well. Tampella's Integrated Boiler Plant Design (IBPD) system indicates the benefits that can be obtained. The reduction in the number of design hours for air ducting for recovery boilers from 2000 hours to 500 hours is one type of significant benefit; the decrease in elapsed time from 16 months to 11.5 months to get working drawings is another.

5.12. TRADEMARKS

Design++ is a trademark of Design Power, Inc. AutoCAD is the registered trademark of Autodesk, Inc, and Knowledge Engineering Environment, KEE, and KEEpictures are the registered trademarks of IntelliCorp. Computervision and Calma are registered trademarks of Prime. VAX is a trademark of Digital Equipment Corporation. Interleaf, Oracle, Sybase, and Symbolics are trademarks of their respective firms. FrameMaker is a registered trademark of Frame Technology Corporation. SPARCstation is a registered trademark of Sun Microsystems.

5.13. ACKNOWLEDGMENTS

Figures 5-1 through 5-11 and Tables 5-1 and 5-2 appeared or are adaptations of material appearing in [5] and are reprinted with the permission of the Center for Integrated Facility Engineering (CIFE). The authors also thank Vaughan Johnson for reviewing the manuscript and suggesting improvements.

5.14. REFERENCES

[1]   Cuppello, J. M. and Mishelevich, D., "Managing Prototype Knowledge/Expert Systems Project," *Communications of the ACM*, Vol. 31, No. , pp. 534-541, .

[2]   Levitt, R., Axworthy, A., and Katajamäki, "Automating Engineering Design with Design++," Nikkei AI Journal, 1991.

[3]   Mishelevich, D. , "Evolution of the Knowledge Systems Market Place: The Intellicorp Experience," in *Managing AI and Expert Systems*, De Salvo, D. and Liebowitz, J. , Ed., Prentice-Hall, Inc. , pp. 80-92 , 1990.

[4]   Riitahuhta, A., *Enhancement of the Boiler Design Process by the Use of Expert System Technology*, unpublished Ph.D. Dissertation, Acta Polytechnica, Scandinavica, Helsinki, 1988.

[5]   Riitahuhta, A., *Enhancement of the Boiler Design Process by the Use of Expert System Technology*, Technical Report 24, CIFE, Dept. Civil Engineering, Stanford University, March 1990.

Chapter 6
KNOWLEDGE-BASED ENGINEERING DESIGN AT XEROX

Lynn C. Heatley and William J. Spear

ABSTRACT

In this chapter we describe three knowledge-based engineering application projects done at Xerox using Wisdom Systems' Concept Modeller^{TM}. Then, we comment on future directions for knowledge-based engineering in Xerox based on what was learned from these projects. The projects have proven the utility and practicality of this new technology in allowing the engineers to build models of complex product subsystems. The models are sufficiently robust to consider design functionality and manufacturing process constraints, and to configure assemblies for major copier subsystems while including standard components. The models are also capable of generating process plans for major sub-system components.

6.1. INTRODUCTION

The driving force behind the work is the business goal to drastically improve the cost-effectiveness of the design/fabrication process. The goal can be broken into three components:

1. Reduce the time required to develop and deliver electro-mechanical systems to the market.

2. Improve quality of our products - consistency, accuracy, and manufacturability. This increases customer satisfaction with our products.

179

3. Reuse and incrementally improve existing designs, sub-systems, components.

In theory, one way to achieve these goals is to build an executable, "intelligent" model of design. These models need to contain more information than the geometry in today's CAD dataset; they should also contain the logic and relationships between design elements that drive the geometry. The logic and relationships should include performance constraints, tolerance analysis, finite element modeling, and manufacturing logic. This model should function as a "living" archive of design and manufacturing knowledge for the product. Xerox recognized that this could be achieved with state of the art of technology.

Considering the recent developments using object technology, graphical user interfaces, and the continued increases in computer performance along with decreasing costs, these questions needed to be answered:

- Is there a computer environment available today that is powerful enough for practical-size models?

- Is the object knowledge representation paradigm robust enough, powerful enough for representing engineering models?

- How hard is it to build such models?

It was determined that the model users needed to be the engineers. But were the tools easy enough to use that engineer users would be capable of building the models in the first place? Or were computer scientists needed to build them?

Commercial tools were surveyed, and projects were launched to answer these questions. The tool chosen for these projects is the Concept Modeller™ from Wisdom Systems. In the next section we will describe the Concept Modeller. This will be followed by a description of various projects at Xerox Corporation and future directions for work at Xerox.

6.2. THE TOOL

Wisdom Systems' Concept Modeller is a commercially available design automation environment. The Concept Modeller is a software toolset that enables companies to implement simultaneous engineering by automating portions of the overall design process for complex products. The toolset allows companies to define and build a complete product architecture definition. The architecture can include engineering criteria, manufacturing process constraints, and commercial considerations that are used in the overall design process. Design standards and organization experience from a range of disciplines can be defined within the product architecture resulting in a consistent, cost- effective, manufacturable product design.

The Concept Modeller provides a way to define product architectures in a hierarchical fashion, i.e. part, sub-part, or sub-sub-part. Component definitions include spatial information, functionality, special properties, and interdependencies with other components. This results in a very tightly coupled product model composed of "intelligent" components. Change effects are propagated throughout an entire product model, when component specifications or substitutions are made, based on the defined interdependencies defined with each part.

As purchased, the environment contains a library of basic geometric and non-geometric primitive objects. These objects are true "classes" of objects in the object-oriented programming style, i.e., functionality associated with object-oriented programming paradigms (multiple class inheritance, class property association, class methods definition, etc.) are implemented in a product definition language. Geometric class primitives are prismatic and cylindrical shapes as well as more advanced geometry such as Non-Uniform Rational B-Spline (NURBS) surfaces.

Non-geometric class primitives are those that might not be created by themselves, but typically are holders and suppliers of information that is applied across multiple product sub-part definitions. A bill-of-materials primitive is an example of a non-geometric primitive, in that it may contain only property information pertinent to "billing out" a sub-part or sub- assembly such as a "part-number." Every part or assembly deemed "billable" in the product definition would inherit from the "bill-of-materials" primitive to gain the property "part-number" and the proper "billing out" procedure. Every part definition developed becomes a primitive class object for the product modeled, and can also be used by other product models (see Figure 6-1).

The way users utilize the Concept Modeller is to modify the primitive objects and relationships to suit a particular sub-system design model. The resulting model can contain all the constraints, rules, and knowledge desired. The knowledge of how to design and manufacture the product is incrementally captured in each part definition. This "how-to" knowledge is captured as property formulas. A part may own as many properties as are required to describe it's behavior in relation to the other parts of the product architecture. Valid property formulas range from a simple constant (a number or symbol) to complex equations, and from an SQL database query to a call to an analysis program. All of these may reference any other part's property values by name as a variable. This mechanism sets up an interdependency network that the Concept Modeller automatically maintains, and allows for rapid "what-if" analysis.

Communication modules are available to access relational database

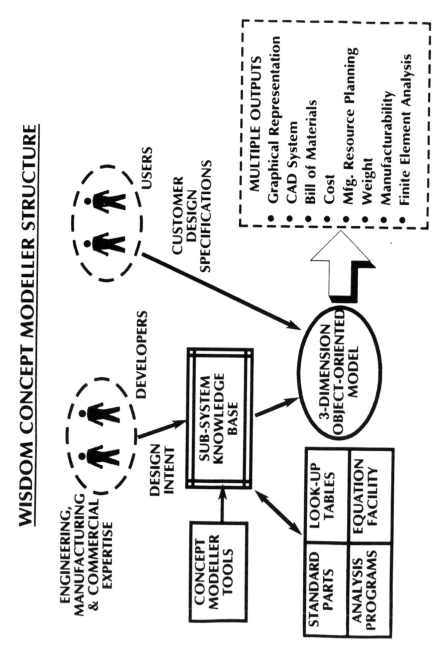

Figure 6-1: Concept Modeller Product Structure

management systems (RDBMS), analysis programs, C or FORTRAN programs, and CAD systems. These modules allow the Concept Modeller to function as the integrator of the CAD system, Bill of Materials system, Finite Element Modeling programs, etc., and allows access to existing standard part and component libraries and cost databases (see Figure 6-2)

Each of the models derived from the product knowledge base models may be shared across a network; not all model information for a particular product needs to reside on a single workstation. This is an important aspect for performance and model data integrity in a large engineering project. If the nature of the product is such that teams of engineers develop various sub-assemblies of the product, a knowledge-based approach can allow these teams to coordinate their design efforts, using private and public (shared) workspaces for each team member, and with configuration control. This requires that the data structures representing the parts (objects) of the product design model and the dependencies between them must be persistent and shared. The Concept Modeller's Project Director module addresses these requirements through object-oriented database management system (OODBMS) functionality.

How the information in the model is presented to the user is as important as the information itself. It must be presented in a way that is easy for the user to understand. The Concept Modeller provides a user interface builder tool that becomes dynamically linked to the product definitions. The user interface builder is based on the industry standards X windows and Motif. Several standard presentation formats are included in the as-purchased system. Users may easily modify the user interface to suit particular viewpoints of the model, such as the manufacturing viewpoint, the cost viewpoint, or the stress analysis viewpoint, by building report panels, icons, and buttons.

6.3. XEROX PROJECTS

Three projects have been undertaken in Xerox using the Concept Modeller: 1) Document Transport Subsystem Project; 2) Scan Subsystem Project; and 3) Fuser Roll Design Project.

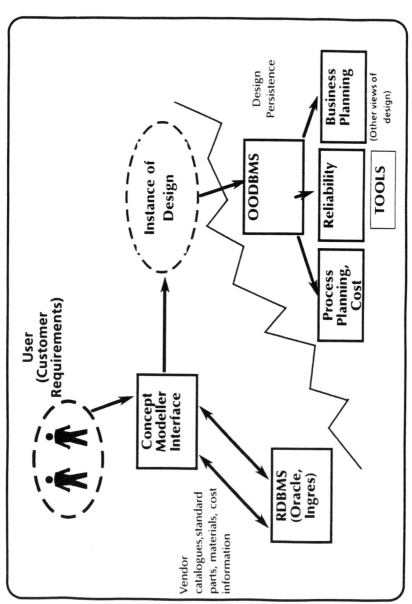

Figure 6-2: System Integration

6.3.1. Document Transport Subsystem Project

Document Transport is an important part of copier design. This is the part of the copier that moves the originals over the platen and returns them to the copier user. The design is driven by geometric constraints. One reason this was chosen as the first problem using the Concept Modeller is that the problem was well understood, based on experience gained by looking at it using other approaches.

On this project, the Concept Modeller was used to solve the problem of placing components comprising a certain kind of document transport system into a defined space. The defined space is a box-like volume. The components are standard components such as rollers, selected from a database of standard components.

The constraints between the components and between the box and the components are geometric. For example, rollers were located within baffles for positioning. A critical design problem for accurate high-speed document transport is placing two rollers relative to each other to maintain the correct pressure and contact area throughout a range of copier operating parameters. Each roller may be composed of more than one material, and dynamic deflections and deformations must be considered. Correct design allows the paper to be driven at high speed thru the sets of rollers without deformation.

From this project it was concluded that the Concept Modeller was well-suited for the purpose of configuring a quantity of known components into a defined volumetric space, while meeting certain geometric constraints. It could be used to successfully layout, configure, and "assemble" a small subsystem, and could successfully capture designs, including component selection strategies based on geometric constraints.

6.3.2. Scan Subsystem Project

The requirement for the second project was to capture more complex design relationships and models. The scan subsystem is one of the most critical in copier design. The design is driven by physical system modeling, design rules and constraints, and manufacturing rules and constraints.

The initial design of the scan subsystem started with defining and satisfying certain customer requirements for the product. These requirements were modeled as forms of constraints that could drive the physical models of the scan system optical components to arrive at a first-try design configuration. The system then iterated thru a series of "standard" component parts in order to arrive at a stable, acceptable optical scan subsystem. In this project and in these design models, manufacturability constraints were included. The "standard" components modeled were not "out of a catalog" standard parts. Our definition of "standard" for this project was expanded to mean those part designs that could be manufactured using known processes. Manufacturing process capabilities were captured as part of the design model. All of the relationships between design components follow the rules put in by the designer of the model.

Figure 6-3 shows part of the hierarchy for the copier and scan system. This hierarchy is a standard part of the Concept Modeller, and is one of the main aids to navigation in the LISP model world. Engineering and manufacturing Bill of Materials generation is obvious and natural from this hierarchy. When building the model, the builder puts it together using a modular approach, extending and expanding the hierarchy at will.

A sample screen state from the Scan-System model is shown in Figure 6-4. The assembly/sub-assembly/part hierarchy is shown in the right-side window. Typical fast-drawn graphics are shown in the large left-side window. Button icons for user interaction are in the lower left window.

- Clicking on "Create-copier" begins creation of a hierarchy for the copier, and adds parts and subparts as needed. This model does not create a model for the entire copier today, just the scan-system. The other subsystems could be added to this model in the future. As the model is exercised to build a specific design, the user is asked to fill in some initial information about the copier design. These high level parameter drivers of scan system design were determined by going through the knowledge engineering process and the model-building process.

- Clicking on "Draw-copier" draws the copier on the screen, as shown in the window above.

- Clicking on "Input-forms" brings up an input form window, in which the design-driving parameters are listed along with an open line for the user to fill in a value.

The lower center window contains the values of several of the parameters. Some of these values may have been determined by the program, and some may have been specified by the user. These values can be brought up by selecting the "Output-forms" icon.

An executable model of the design greatly facilitates what-if analysis. This is done by simply changing some of the input parameter values and propagating them through the model. Figure 6-5 shows the difference in potential document outline between "long-edge" feed direction and the "all" feed direction. We can encode into the model what it means to the design for the user to select "long-edge" or "all."

This project successfully demonstrated the ability to capture more complex models using knowledge-based engineering tools.

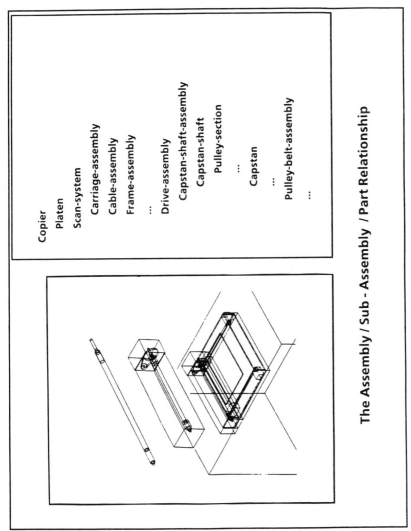

Copier
Platen
Scan-system
   Carriage-assembly
   Cable-assembly
   Frame-assembly
   ...
   Drive-assembly
      Capstan-shaft-assembly
         Capstan-shaft
         Pulley-section
         ...
         Capstan
         ...
         Pulley-belt-assembly
         ...

The Assembly / Sub - Assembly / Part Relationship

Figure 6-3: The Assembly/Sub-assembly/Part Relationship

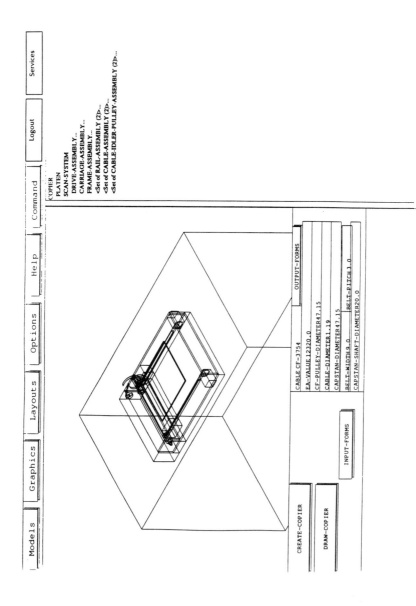

COPIER
PLATEN
SCAN-SYSTEM
DRIVE-ASSEMBLY...
CARRIAGE-ASSEMBLY...
FRAME-ASSEMBLY...
<Set of RAIL-ASSEMBLY (2)>...
<Set of CABLE-ASSEMBLY (2)>...
<Set of CABLE-IDLER-PULLEY-ASSEMBLY (2)>...

Models   Graphics   Layouts   Options   Help   Command   Logout   Services

CREATE-COPIER

DRAW-COPIER

INPUT-FORMS

OUTPUT-FORMS

CABLE CF-3754
EA-VALUE 12320.0
CF-PULLEY-DIAMETER 47.15
CABLE-DIAMETER 1.19
CAPSTAN-DIAMETER 47.15
BELT-WIDTH 9.0   BELT-PITCH 1.0
CAPSTAN-SHAFT-DIAMETER 20.0

Figure 6-4: A Screen from the Scan-System Model

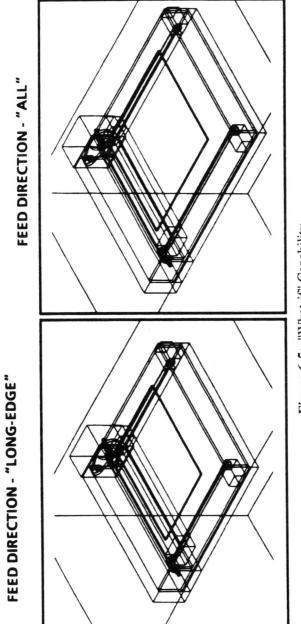

Figure 6-5: "What-if" Capability

6.3.3. Fuser Roll Design Project

The third project is fuser roll design. The fuser roll is a roller in the copier that uses heat and pressure under very precise conditions to fuse the toner particles to the blank copy paper, thus producing the copy. Copy quality is determined very directly by the design of the fuser roll. Design of the fuser roll is driven by fuser technology process requirements, physical system modeling, design rules and constraints, and fabrication rules and constraints. The degree of sophistication and complexity increased significantly from the scan system project.

The fuser design project was driven by the manufacturing community. Fuser roll manufacturing involves inertia welding and machining. In the past, for different fuser roll configurations, manufacturing processes and the parameter values were not consistent or reliably predictable by the engineers. Process try-out and modification was needed. The final correct values were recorded, but a model of the process solution-space was not built. When another fuser design needed to be manufactured, process try-out was needed again. The goal with the Concept Modeller was to help understand the process parameter space and then build a process model. The Concept Modeller was used to help organize the knowledge as it was gained. This was a major difference from the previous projects where the design model was understood before beginning the project.

After reviewing the knowledge and information requirements for the project, it became obvious that the manufacturing group needed access to more information to define the processes. They also needed expertise in fuser roll design. This expertise and information existed in the Technology and Engineering Departments. Consequently, the project and the model were expanded from fuser roll process design to fusing design in general, using information and knowledge from the engineers in the Technology and Engineering Departments.

A screen state from the Fuser Roll project is shown in Figure 6-6. The right window shows the assembly/sub-assembly/part relationship. Notice that the fuser roll definition comprises a fabrication component, a core-assembly component, and an elastomer component, and that the fabrication component has a machining view. This allows us to concentrate the machining factors in a separate area of the model for easy understanding. Notice the pressure roll, which works directly with the fuser roll to fuse the toner to the paper. The three dots after some entries indicates additional levels of hierarchy are present but not displayed. The center window shows the fuser roll before, during, and after fabrication and attachment of the end caps, and shows the fuser roll and pressure roll in their as-designed working relationship as determined by the forces between them. The upper left window shows the user's choices. This window and these choices were programmed by the engineer as a customized user interface, within the Concept Modeller. Notice that "Engineer Fuser Roll" and "Engineer

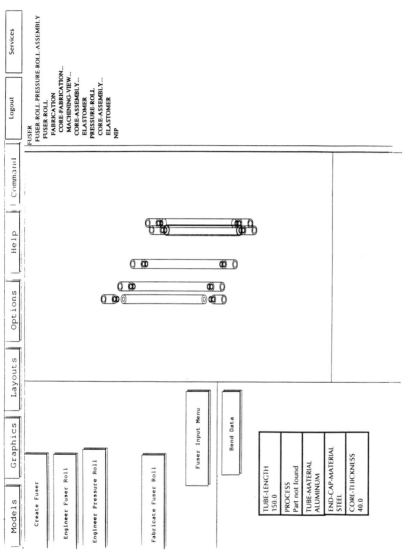

Figure 6-6: A Screen from the Fuser Roll Project

Pressure Roll" are separate icons. This allows the user to begin with either one, depending on what is known and what the user wants to use as dependent variables. "Fabricate Fuser Roll" results in a process plan with process parameters determined. "Fuser Input Menu" provides a standard input form for the user. The lower left window reports some values to the user. "Bend data" brings up a different report.

The model now encompasses physical system design modeling, fusing technology requirements, and process requirements. Once these requirements are satisfied, the components are assembled by the modeller using various engineering rules and constraints, including selecting materials based on their properties. From this point, the generative process planner function executes, in which the fabrication process steps are generated by acting on the model, based on manufacturing rules and constraints.

This project demonstrated that manufacturing processes could be captured along with the design drivers.

6.4. QUESTIONS ANSWERED, LESSONS LEARNED

6.4.1. Questions Answered

Three projects were undertaken in order to answer some questions in light of the recent developments using object technology, graphical user interfaces, and continuing increases in computer performance along with decreasing costs. These questions can now be answered:

1. **Question:** Is there a computer environment available today that is powerful enough for practical-size models?
   Answer: Yes. The high-powered workstation is powerful enough. Lots of memory and disk space make executing large LISP worlds practical. Even so, operations such as designing and assembling a subsystem can take a while. Still, compared to doing the same operations manually, it is faster by several orders of magnitude. It is now repeatable, and all aspects of the design that are included in the model are automatically checked. Manually it can take days to perpetrate a change through the design model, and this time is often the limiting factor on how many "what-if" analyses will be done. Using the computer, changes can perpetrate thru the model and users can look at the results in minutes or hours where days or weeks were needed before.

2. **Question:** Is the object knowledge representation paradigm robust enough, powerful enough for representing engineering models?
Answer: Yes. Objects and methods were sufficient to represent the models and problems.

3. **Question:** How hard is it to build such models?
Answer: It requires training and time working with the system, similar in length to that for using a solids and surfaces CAD system. During that time, the user also develops a basic working knowledge of LISP. It takes about about as much time to build the model as it takes to create the first design manually. However, each major design iteration requires days or weeks manually, and iterations were done in minutes or hours using the computer model. In the case of the fuser roll project however, a model probably wouldn't have been built at all, without the single computing environment provided by the Concept Modeller for both the design constraints and for the manufacturing process model.

4. **Question:** It was determined that the model users needed to be the engineers. But were the tools easy enough to use that engineer users would be capable of building the models in the first place?
Answer: Yes, given the training and the time to work with the system as described above.

5. **Question:** Or were computer scientists needed to build them?
Answer: Engineers with computer science training were used on these projects, with good results. This mix of skills is preferred.

6.4.2. Lessons Learned

Design models and manufacturing process models can be successfully captured using this technology. The Knowledge-based Engineering model-building process results in an executable model of the design, and so it requires dependencies to be stated explicitly. This takes work, but helps identify conflicting requirements and facilitates documenting "why." The Knowledge-based Engineering environment is like a two-edged sword: very powerful, but not easy to use at first. There is a long training period like with CAD systems. There was no seamless integration with CAD system geometry until recently. Software costs are high. The "capture" of knowledge and building the interfaces is time-consuming. However, it provides a record of design intent, decisions, and reasons for design details that can be passed along to the next engineers who work on the project. It takes as long to build the each sub-system model initially as it does to design the sub-system manually. There is an advantage, though, in that a partially completed model is useful, where a partially designed subsystem may not. The resulting models allow us to do faster, and therefore, more design iterations and design modifications. Use of Knowledge-based Engineering models results in less new part designs and more standardization. The resulting designs can have built-in manufacturability, which means higher quality designs. In all, Knowledge-based Engineering projects can be worth the effort and the cost.

6.5. FUTURE DIRECTIONS

The driving force behind this work was the business goal to drastically improve the cost-effectiveness of the design/fabrication process. The goal was broken into three components:

1. Reduce the time required to develop and deliver electro-mechanical systems to the market.

2. Improve quality of the products - consistency, accuracy, manufacturability, thus increasing customer satisfaction.

3. Reuse and incrementally improve existing designs, sub-systems, components.

Here is a vision of the design process when nearing the goal: Designers, engineers, manufacturing engineers, and cost engineers use a common underlying computer-based format for product specification information. There will exist a unified conceptual and computational model of each design. This model is stored in a distributed manner among the machines of the various engineers. They work concurrently on the common model of the design and on models of manufacturing processes. Each user is in control of which information is presented on his or her screen and how it is presented - each user can have his or her own application viewpoint.

An approach to achieving the vision: Use knowledge-based systems technology to capture Xerox-specific knowledge of subsystem design and manufacturing process capability Each sub-system knowledge-based model consists of the best expertise obtained from combined input of technology, design, manufacturing, and service engineers.

6.6. SUMMARY

Three projects were built using Wisdom Systems' Concept Modeller, with increasing complexity in each successive project. The projects have proven the utility and practicality of this new technology in allowing the engineers to build models of complex product subsystems. The models are sufficiently robust to consider design functionality and manufacturing process constraints, and to configure assemblies for major copier subsystems while including standard components. The models are also capable of generating process plans for major subsystem components.

Larger-scale implementation of knowledge-based systems technology will result in significant progress toward all three components of the business goal:

1. Reduce the time required to develop and deliver electromechanical systems to the market.

2. Improve quality of the products - consistency, accuracy, and manufacturability.

3. Reuse and incrementally improve existing designs, sub-systems, and components.

Xerox is embarking on more KBE projects, training more engineers. The incorporation of this technology into the product design and manufacturing process is viewed as critical to success in the 1990s.

PART VIII: INTEGRATED ENVIRONMENTS

Chapter 7
AN INTELLIGENT CAD SYSTEM
FOR MECHANICAL DESIGN

Nien-Hua Chao

ABSTRACT

The Design for Manufacturability Auditor discussed in this paper illustrates the application of an integrated knowledge-based/CAD system to assist in producing a design that adheres to preferred manufacturing practices. This effort is but one step in the long journey toward the development of intelligent CAD systems for mechanical design. Our objective is to develop a CAD system that consists of features that assist users in following the DFX ("Design For X") practices in an early product development stage. The "X" in "DFX" stands for "manufacturability, assembly, repairability, etc."

The development of intelligent CAD systems for mechanical design is often interrupted by obstacles. Some of the issues confronting today's researchers include design representation, managing and using existing components and subsystems, geometric reasoning, knowledge-based systems, and open architecture of vendor tools.

However, new algorithms and rule-based system development tools are steadily being developed to apply to advanced knowledge-based systems, user-programmable geometric modeling systems for implementing geometric reasoning, and data abstraction for programming in an object-oriented style. These new tools will dramatically alter the mechanical design process: both the development of future computer-aided design systems and the way a designer uses a CAD system to develop a new product.

199

7.1. INTRODUCTION

To succeed in today's competitive world market, product developers are under tremendous pressure to shorten the product development cycle and to improve product quality. In the design process, we rely heavily on computers. Computer-aided engineering (CAE), computer-aided design (CAD), and computer-aided manufacturing (CAM) systems improve and expedite the mechanical product design and manufacturing process. In general, designers depend on CAD systems that were developed using the technology of the 1970's. In the current generation of CAD systems, however, product information is prepared for human designers to read: e.g., the tolerance of a line is presented as text, not as an attribute. The following typical CAE/CAD/CAM tools are useful:

- Finite-element programs [2] analyze stress and thermal patterns in designs.

- Optimization algorithms [10, 30] help designers achieve optimal designs.

- Solid modeling packages [5, 21, 31] generate model images in the conceptual design stage.

- Computer-aided drafting systems [6, 7, 38] prepare detailed design and design documentation.

- Numerical control (NC) machining modules [8, 19, 39] simulate and verify the cutter path and then generate NC machining process instructions.

To use these tools effectively in product design, we need well-trained users. In the research community, scientists and engineers are developing new tools [3, 13, 16] [17, 20, 36, 40] and algorithms that will dramatically alter the direction of new computer-aided design systems and the usage of CAD systems in developing new products. These new tools and algorithms include tools for developing rule-based systems, user-programmable geometric modeling systems for implementing geometric reasoning, data abstraction for programming with object-oriented style, etc.

A major concern of CAD system developers is that most design information for existing components and subsystems is stored as blueprints. Compared to the pioneering efforts at using CAE/CAD/CAM systems for design, too little effort has been placed on cataloging and computerizing design information. Developers need to easily retrieve and use design information for new product development.

What are other areas in which we can either foresee problems or propose improvements for the design process described above? First, we may not have enough well-trained personnel to use CAE/CAD/CAM systems effectively. Secondly, we need to develop a systematic approach to converting existing component and subsystem information from a paper format to a digital format. This change will improve system management and utilization. Third, we need to develop a methodology to organize design information generated by CAE/CAD/CAM systems so that a product developer can query and retrieve this information to reuse in new product designs. Finally, design reviews may not be followed properly because either the guideline is obsolete or the designer just does not check it. These are DFX issues.

This paper discusses the importance of the product design process in the product development cycle and describes an effort to develop an intelligent computer-aided design system for mechanical product design to solve some of the current problems.

7.2. MOTIVATION -- THE IMPACT OF THE DESIGN PROCESS

A sound initial design guarantees manufacturing flexibility: freedom to select appropriate materials, manufacturing methods, and the product assembly procedure. A poor initial design, on the other hand, usually results in a poor quality product because a designer balks at making any major design modifications unless a major design deficiency becomes evident.

Unless a designer considers functional requirements along with the manufacturing methodology of the product, the resulting unnecessary constraints on product development will indirectly affect product quality and manufacturing costs. For example, the proposed design may limit materials used, manufacturing methods and processes, methods of product assembly, and product maintenance procedures.

Costs of manufacturing directly affect a product's success or failure in today's competitive world market. The major portion of product manufacturing expenses are determined in the early design stages. Typically, 50% to 75% of the total manufacturing cost of a product is determined during the preliminary design. The remaining 25% to 50% of manufacturing costs are determined by production setup, processing (manufacturing, assembly, inspection, etc.), purchasing strategies, and inventory control.

To gain a competitive edge, product developers are highly motivated to find ways to trim expenses. Current efforts to improve the competitive edge have focused on shortening the product development cycle, optimizing the manufacturing processes, and improving product quality. Product developers are exploring more effective ways to use computer-aided design tools in the product development process.

7.3. DEVELOPING AN INTELLIGENT CAD SYSTEM FOR MECHANICAL DESIGN

Mechanical product development is a slowly evolving science, which is profoundly impacting the manufacturing process. At present, there are many research issues [1, 9, 15, 18, 24, 23, 26], [29, 34, 33] in developing an intelligent CAD system for mechanical design. In this section, we will first discuss a simplified mechanical product development process and how it evolved with the advancement of computer technologies; then, we discuss some research issues in developing an intelligent CAD system.

7.3.1. Mechanical Design Process

The simplified mechanical product development process shown in Figure 7-1 consists of five stages. Starting with a design intent, a designer searches through existing components and subsystems to locate those that will satisfy the design needs. With this information, the designer must decide if the existing design should be modified or redone entirely. After completing the modification or preliminary design, a designer should analyze and simulate the proposed design to determine if the proposed design will perform as required. The next step is to review the design with manufacturing process planning, production engineers, etc. Finally, the design is ready for prototyping.

Advancing computer technology is rapidly changing the use of computers and computer-aided design tools in the product development process. As a result, the mechanical product development process is becoming more sophisticated as it subsumes these new principles (Figure 7-2). Advancement in the following areas has a significant impact on the product development process:

- Migrating design information from blueprints to a data base management system

- Developing on-line DFX auditors to accommodate design review meetings

- Developing knowledge-based systems as front-end processes to ensure product quality. These new systems can assist inexperienced designers to produce expert-level designs.

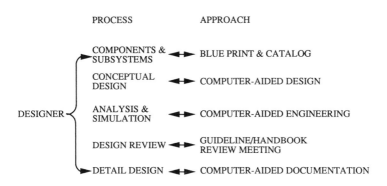

Figure 7-1: Mechanical Product Development Process - Current

7.3.2. Research Issues in Developing an Intelligent CAD System

To integrate these design stages described above into an intelligent CAD system, we are studying the following issues:

- Design representation
- Management and use of existing components and subsystems
- Geometric reasoning
- Knowledge-based systems
- Open architecture

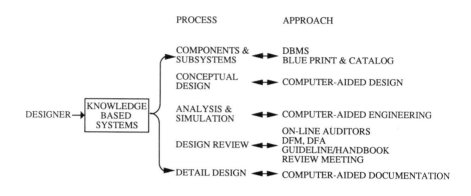

Figure 7-2: Mechanical Product Development Process - Evolving

7.3.3. Design Representation

The mechanical product development process and the product information associated with it are highly complex. Typical mechanical product data consist of information for the following product realization processes:

- Conceptual and detailed design
- Product performance analysis and simulation
- Manufacturing process planning
- Specifications for product manufacturing
- Product assembly

Because a designer does not require all of the product information at any one particular stage of the design process, we are looking for a proper methodology to model product information. Figure 7-3 represents a typical design prepared by an existing CAD system. Much of the design information in this drawing is prepared for humans to read and to understand: the tolerance of an entity and the engineering notes are prepared as text, not as an attribute of the entities. A computer cannot automatically interpret this design information without ambiguity.

The research issue is to define a product data model that can be correctly interpreted and manipulated by a computer. After this product data model is defined, we then need to define a methodology to organize this information. The information needs to be in a form that can be used effectively in the various stages of the product realization process. For example, product data that consist of geometry, material property, thermal, tolerance, surface finish, and electronic characteristics of a design are not all used at any single design stage. However, this compendium of information needs to be part of a product data model that can be accessed at different design stages to fulfill multiple design requirements.

7.3.4. Managing and Using Existing Components and Subsystems

An experienced (expert) designer knows how to effectively reuse existing components and subsystems to design a new product. An experienced designer's ability to remember past designs and reuse them in new, innovative designs is a very important consideration in the development of an intelligent CAD system. The issue, then, is how to store information of existing components and subsystems in an electronic format for query and retrieval. The problems are manifold. For example,

- Which existing components or subsystems should we store in any particular design data management system (the functionality of the designs, physical size of the designs, etc.)?

- How can we effectively organize this information so that it can be queried and then reused for designing new products?

- What kind of data management system should we use? Should we retrofit a commercial data base management system such as Oracle [27] or INGRES [22], that was developed for business applications, or should we develop a new one for engineering applications?

- How can we develop a scheme to effectively keep the design data base updated? We need to consider how to maintain an up-to-the-minute design data base.

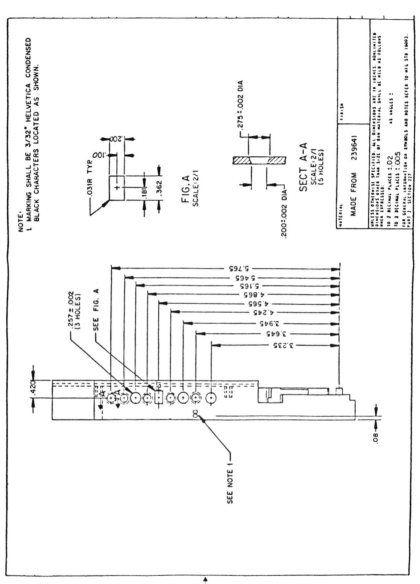

Figure 7-3: A Drawing of a Design Prepared for a Human Being to Read

7.3.5. Geometric Reasoning

A product with high quality and cost-effective manufacturing is the key to competing successfully in this worldwide market place; one way to achieve this goal is to apply DFX rules in product development process. Currently, human product developers are the key in applying the DFX rules in the product development process. How can we use computer aids to implement the DFX rules in the product development process?

Geometric reasoning holds the key to implementing a computerized enforcement of DFX in the product development process. Currently, though, an experienced designer can perform geometric reasoning in a product design process more quickly and more easily than a computer. Geometric reasoning allows us to uncover design flaws. For example, geometric reasoning would point out the weakness in a cylinder and hollow tube junction that is exposed to high stress. The expert designer could suggest adding a fillet between these two elements to reduce the stress concentration and allow the joint to endure. To mimic this human design reasoning process, we are currently using a user-programmable solid modeler [28] to study this issue. By using the solid modeler to design a cylinder and hollow tube junction, we first call functions to create and then join the two elements together, as shown in Figure 7-4. From experience, we know that a fillet is needed between these two elements. We can then call another function to create a fillet, shown in Figure 7-5. Eventually, this human design expertise and decision process should be incorporated into a knowledge-based system that can be used to create this design automatically.

If a two-dimensional picture is worth a thousand words to a designer, a three-dimensional solid model is worth many times that amount. A three-dimensional solid model conveys the design intent to the user without ambiguity. Therefore, an intelligent CAD system that can communicate with a user and display a solid model of the design intent is an extremely valuable tool. With a user-programmable modeler, we can develop both a message-driven and menu-driven user interface. The message-driven feature of the system allows knowledge-based systems to exchange design ideas with its users through three-dimensional graphics. The menu-driven feature of the system allows an experienced user to modify a design as desired.

7.3.6. Knowledge-Based Systems

In developing an intelligent CAD system, we need to capture human design expertise [35, 37] and implement it as a set of knowledge-based computer-aided systems. High quality products and shortened product development cycles are

Figure 7-4: A New Part is Created by Joining Two Elements

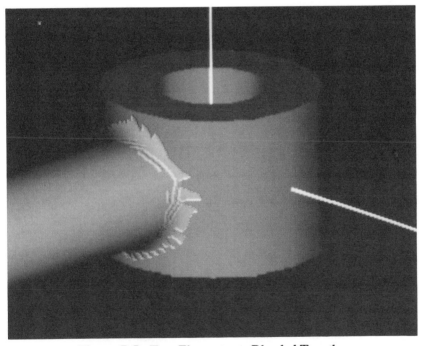

Figure 7-5: Two Elements are Blended Together

two key elements in improving a company's competitive edge. A complication in reaching this goal is that product development experts are in big demand but in short supply. Experts are invaluable because they can:

- effectively reduce a complex problem to a set of subproblems identify critical parameters interpret analytical results and properly adjust design variables

We need to capture human design expertise first; then we need to select a tool for developing knowledge-based systems. What are the issues that relate to the selection of a vendor tool and/or the creation of an in-house developed knowledge-based system development tool? Can the tool:

- be integrated with other computer-aided mechanical design software packages?

- accommodate expertise from multiple sources? Mechanical design, for example, is a multi-disciplinary activity, which includes thermal, structural, kinematics, dynamics etc.

- coordinate expertise from multiple sources? Competitive design plans emerge from the joint effort of a group of designers. Negotiation and trade-offs typify this product development.

7.3.7. Open Architecture

Open architecture is an important criterion in selecting vendor tools. Using vendor-supplied general purpose CAD tools gives us a cost advantage because vendors spread development costs over a broad customer base. The risk, however, of depending on vendor tools is that vendors are not always willing to upgrade and improve obsolete tools. By selecting a vendor tool with open architecture, we can minimize this risk.

Vendor tools with open architecture offer the following advantages:

- Design information generated by using vendor tools can be easily accessed.

- Vendor tool design information can be modified without difficulty

- Vendor tools integrate easily with other tools and they easily exchange information between modules

- We can always integrate in-house developed tools with appropriate vendor tools. This desired capability preserves in-house development investments.

Although we still have a number of research issues in achieving a truly intelligent CAD system for mechanical design, we have made great strides in reaching our goal. In the next section, we will describe a knowledge-based system to illustrate one development effort.

7.4. EXAMPLE -- A DESIGN FOR MANUFACTURABILITY AUDITOR

In this section, we will illustrate how to embed human design and manufacturing expertise into a knowledge-based Design for Manufacturability Auditor (DFM Auditor) [11] and how to integrate this DFM into a computer-aided design system. The results will allow designers to create higher quality products.

7.4.1. Design for Manufacturability

What is design for manufacturability? A product can be effectively manufactured by using available equipment. Some product features that are difficult to manufacture using a machining process are:

- Incomplete or incorrect designs
- Sharp corners
- Tight tolerances
- Nonstandard hole sizes
- Too many widely varied hole sizes

These design anomalies increase manufacturing costs, affect product quality, and therefore lower the competitive position. A DFM auditor that is coupled with a general purpose CAD system can help designers catch these and other design problems before the design is ready to manufacture.

7.4.2. Modules of the DFM Auditor

The DFM Auditor consists of three modules, shown in Figure 7-6. The CAD system interface module extracts product design information from different CAD systems. The two foundation modules, located in the middle right and left of the figure, provide knowledge and expertise. The third module provides recommendations of designs that conform to manufacturing practices. The DFM auditor receives a CAD design file, interprets the design information, evaluates the design, and makes recommendations based on design for manufacturability practices.

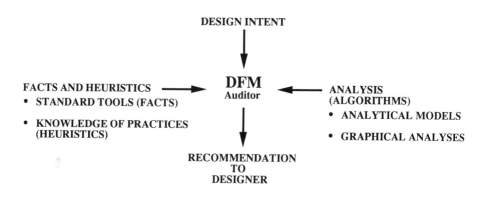

Figure 7-6: Structure of the Design for Manufacturability Advisor

7.4.3. CAD System Interface and Data Structure

The DFM Auditor analyzes design files from different CAD systems by converting CAD generated design data files to an industry-standard file format, e.g. IGES/PDES [4, 14, 32]. Then it submits the files to the DFM auditor.

The CAD system interface module of the DFM auditor first digests the information and discards unrelated information from the design file. The unrelated information includes dimensions of lines, circles, etc. and document fonts. Then, the DFM auditor reconstructs the design features [11, 12, 25]. An example of a simple design feature is a tapered hole, shown in Figure 7-7. The CAD system represents the tapered hole by using two circles and one line.

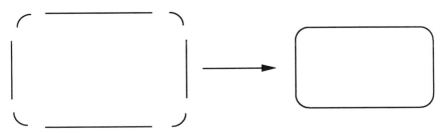

Figure 7-7: Four Lines and Four Arcs Construct a Rectangle with Fillets

In developing the DFM auditor, the attribute-value element, which is similar to the frame structure [40], is created to store the entity information (Figure 7-8).

The entity information associated with a line includes the x, y, z coordinates of the two end points. The attribute-value elements associated with the entity information are listed below:

```
(Entity-value
    entity-number
    1st-field
    2nd-field
    3rd-field
    4th-field
        .
        .
    nth-field
)
```

After receiving the entity information, the DFM auditor reconstructs the design features. Attribute-value elements associated with the stored feature information are listed below and are illustrated in Figure 7-9.

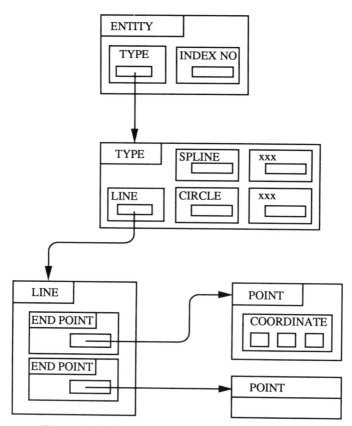

Figure 7-8: Data Structure of Entity Information

```
(Entity
    entity-number
    entity-type
    total-data-point
    radius
    entity-group
    connecting-to
    connecting-from
)
```

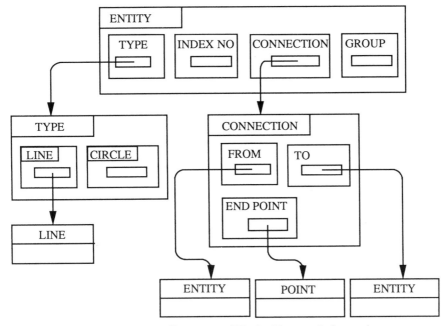

Figure 7-9: Data Structure of Entity Feature Information

7.4.4. Knowledge and Expertise

To create a DFM Auditor, we needed to glean preferred manufacturing practices and expertise from experts. Typical preferred manufacturing practices include: 1) Minimization of the number of hole sizes. This practice reduces tool changes. 2) Specified standard hole size. This practice eliminates special tooling to produce nonstandard hole sizes. 3) Manufacturing site considerations based on capability.

The DFM auditor can access: 1) standard drill sizes stocked by most machine shops and 2) particular drill sizes available or not available at a specific manufacturing location. This information is stored in a table (Table 7-1) or in a data base management system.

Machine Tool Information		
Type	Diameter	Manufacturing Location
drill	0.125	
drill	0.1406	Site A
drill	0.1563	
drill	0.1875	
drill	0.2031	Site B
drill	0.25	
drill	0.3125	
drill	0.3750	

Table 7-1: Machine Tool Information

7.4.5. Production Rules

The production rules stored in the knowledge base of the DFM auditor state preferred manufacturing practices. This implementation helps manufacturers to minimize product manufacturing costs by following DFM auditor's recommendations. One production rule contained in the DFM Auditor is listed in Figure 7-10.

Rule f-rule-20 checks a hole that may require a special tool and, as a result, may cause manufacturing problems. A small tool, for example, can easily break in the machining process.

```
              OPS5 language                    English "If Then" Translation

(p f-rule-20                              ;        If
  { (Processing-entity ^number <x0>       ;        process counter = <x0>,
                 ^diagnosis <no>)<label-1>};
  { (D-status ^diagnosis check-size )     ;        diagnosis status =
                        <label-2>}        ;        check hole size,
  (Entity ^entity-number {= <x0>}         ;        entity <x0> has
      ^entity-type <name> ^radius <r0>    ;        radius = <r0>,
      ^connecting-from <yy1>              ;
      ^connecting-to <yy2>)               ;
  (DFM-criterion ^lower-bound {> <r0>})   ;        and <r0> is too small
  (Entity-value ^entity-number {= <x0>}   ;
          ^1st <x1> ^2nd <y1> ^3rd <z1>)  ;
  -->                                     ;        Then
  (bind <no> (compute <no> + 1 ))         ;        increase message
  (write (crlf)                           ;        counter by one,
         (crlf)                           ;        print warning message
         (crlf) |*** Suggestion number  |;        with entity
              <no> |    *** |             ;        position,
         (crlf)                           ;        increase process
         (crlf))                          ;        counter by one,

  (write (crlf) | The | <name>            ;        and set diagnosis status
              |, whose center is at |     ;        = yes
         (crlf) | |                       ;
         (crlf) | ( | <x1> |, | <y1>      ;
              | , | <z1> | ) |            ;
         (crlf) | |                       ;
         (crlf) | radius = | <r0>         ;
         (crlf) | |                       ;
         (crlf) |   will create a|        ;
              | manufacturing |           ;
              | problem ( too small )|)   ;
  (bind <x0> (compute <x0> + 1 ))         ;
  (modify <label-1> ^number <x0>          ;
                     ^diagnosis <no>)     ;
  (modify <label-2> ^diagnosis yes)       ;
)                                         ;
```

Figure 7-10: A rule in the DFM auditor

7.4.6. Example

A simple design was shown to illustrate the application of the DFM Auditor. The design includes two questionable design features: a rectangular hole with very small fillets and a hole that would need special tooling. After the design was submitted to the Auditor, the Auditor printed out the applicable suggestions, listed in Figure 7-11.

Because the human design expert knows whether or not a specific functional

```
        *** Suggestion number   1    ***

The  Rectangular-hole, whose center is at

(  0.181,   0.1,   0.0  )

has corner radii =  0.031

will create a manufacturing problem ( too small ).


   *** Suggestion number   2   ***

The  Circle, whose center is at

(  -0.9265,  0.0933,  0.0  )

radius =  0.1285  requires an unusual size tool.

Can you modify the  Circle  radius to

   0.125    or  0.1328
```

Figure 7-11: Suggestions Generated by DFM Auditor

requirement of a design necessitates a nonstandard feature, the DFM Auditor must allow the designer to intervene in any design changes. Therefore, the Auditor only provides recommendations; it does not automatically change any feature. The designer can then evaluate the validity of the DFM Auditor recommendation and approve or disapprove of the recommended feature change.

7.5. CONCLUSION

The development of an intelligent CAD system is an evolving process. As researchers developing intelligent CAD systems by mimicking the human expert design decision processes, they often open up many new issues as they solve nagging questions. We are currently probing the following issues:

- How to advance from the current computer-aided documentation environment to a computer-aided design environment

- How to refine both design and manufacturing expertise and incorporate it into a knowledge-based system

- How to provide a flexible design environment by integrating in-house developed knowledge-based systems with a state-of-the-art vendor CAD system

- How to develop design for manufacturability and design for assembly systems as front-end processors for mechanical CAD systems

- How to continuously incorporate new research findings into the development of an intelligent CAD system for mechanical design.

The Design for Manufacturability Auditor represents a new approach in developing a computer-aided design system. The incorporation of the DFM auditor with a vendor CAD system allows designers to create improved designs for products that require a machining process and offers many advantages over traditional design methods: designs are easily produced; they adhere to approved design for manufacturability practices; and, as a result, manufacturing costs are cut.

Intelligent CAD systems are the wave of the future, and will open up a whole new era for product development. Intelligent CAD systems hold the key to fast-paced, improved design development. Their flexibility and potential make them adaptable to the demands of the market place.

7.6. ACKNOWLEDGEMENTS

The author wishes to express his appreciation for the suggestions provided by his colleagues T. J Kowalski, R. W. Lunt and G. T. Vesonder.

7.7. REFERENCES

[1]   Ackland, B., "CADRE - A System of Cooperating VLSI Design Experts," *Proceedings of ICCD 85*, pp. 99-104, 1985.

[2]   *ANSYS - Engineering Analysis User's Manual*, Swanson Analysis Systems, Inc., 1991.

[3]   Barr, A. and Feigenbaum, E. A., *The Handbook of Artificial Intelligence*, William Kaufmann, Inc., 1981.

[4]   Bradford, S. et al, *Initial Graphical Exchanges Specification, Version 2.0*, National Bureau of Standard, 1982.

[5]   Brown, C. M., "PADL-2 A Technical Summary," *IEEE Computer Graphics and Applications*, Vol. 1982, March 1982.

[6]   *CADDS User Guide*, ComputerVision Corp., 1991.

[7]   *CADKEY User Reference Guide*, Micro Control System, Inc., 1991.

[8]   *CAM - FactoryVision*, ComputerVision Corp., 1991.

[9]   Cammarata, S. J. and Melkanoff, M. A., *An Interactive Data Dictionary Facility for CAD/CAM Data Bases*, Expert Database Systems, The Benjamin/Cummings Publishing Company, 1986.

[10]   Chao, N. H., "Application of a Reduced Quadratic Programming Technique to Optimal Structural Design," *Proceedings of the International Symposium on Optimum Structural Design*, Tucson, Arizona, October 1981.

[11]   Chao, Nien-Hua, "The Application of a Knowledge-Based System to Design for Manufacturing," *1985 IEEE International Conference on Robotics and Automation*, St. Louis, Missouri, March 1985.

[12]   Chao, Nien-Hua and Schiebel, E. N., "Inspection Assistant - A Knowledge-Based System for Piece Part Inspection," *1986 IEEE International Conference on Robotics and Automation*, San Francisco, California, 1986.

[13]   Cohen, P. R. and Feigenbaum, E. A., *The Handbook of Artificial Intelligence*, William Kaufmann, Inc, 1982.

[14]   Downey, P. J. et al, *Product Definition Data Interface - Needs Analysis Document*, AFWAL/MLTC, Wright-Patterson Air Force Base, Ohio, July 1983.

[15] Fenves, S. J. and Garrett Jr., J. H., "Knowledge Based Standards Processing," *Artificial Intelligence,* Vol. 1, No. 1, 1986.

[16] Forgy, C. L., *The OPS5 User's Manual,* Technical Report, Department of Computer Science, Carnegie-Mellon University, Pittsburgh, Pennsylvania, 1981.

[17] *The OPS83 User's Manual and Report,* Production Systems Technologies, Inc., Pittsburgh, Pennsylvania, 1985.

[18] Fox, M. S. and McDermott, J., *The Role of Databases in Knowledge-Based Systems,* Technical Report CMU-Ri-TR-86-3, Carnegie-Mellon University, February 1986.

[19] "GMACH Operational Description," 1991.

[20] Hayes-Roth, F., Waterman, D. A. and Lenat, D. B., *Building Expert Systems,* Addison-Wesley Publishing Company, 1983.

[21] *I-DEAS User Guide,* Structural Dynamics Research Corporation, 1991.

[22] *INGRES User's Manual,* Relational Technology, Alameda, California, 1991.

[23] Katz, R. H., *Information Management for Engineering Design,* Springer-Verlag, 1985.

[24] Kowalski, T. J. and Thomas, D. E., "The VLSI Design Automation Assistant: An IBM System/370 Design," *IEEE Design & Testing,* February 1984.

[25] Luby, S. C., Dixon, J. R. and Simmons, M. K., "Creating and Using a Features Data Base," *Computers in Mechanical Engineering,* November 1986.

[26] Mayer, A. and Lu, S C-Y, "An AI-Based Approach for the Integration of Multiple Sources of Knowledge to Aided Engineering Design," *ASME Journal of Mechanisms, Transmissions, and AUtomation in Design,* Vol. 110, No. 3, pp. 316-323, 1988.

[27] *Oracle User's Manual,* Oracle Corporation, Menlo Park, California, 1991.

[28] *Prodigee Geometry Processing Environment - User Manual,* XOX Corporation, Ithaca, New York, 1987.

[29] Rehak, D. R. and Howard, H. C., "Interfacing Expert Systems with Design Databases in Integrated CAD Systems," *Computer-Aided Design,* Vol. 17, No. 9, 1985.

[30]   Reklaitis, G. V., Ravindran, A. and Ragsdell, K. M., *Engineering Optimization Methods and Applications,* John Wiley & Sons, Inc., 1983.

[31]   Requicha, A. A. and Voelcker, H. B., "Solid Modeling: A Historical Summary and Contemporary Assessment," *IEEE Computer Graphics and Applications,* March 1982.

[32]   Smith, B., *Initial Graphics Exchange Specification - PDES Initiation Activities,* National Bureau of Standard, 1986.

[33]   Sriram, D., Maher, M. L. and Fenves, S. J., "Knowledge-Based Systems in Structural Design," *Computer & Structures,* Vol. 20, No. 1-3, 1985.

[34]   Staley, S. M. and Anderson, D. C., "Functional Specification for CAD databases," *Computer-Aided Design,* Vol. 18, No. 3, 1986.

[35]   Strasser, Federico, "Keeping Machining Costs Down," *Machine Design,* August 1982.

[36]   Stroustrup, Bjarne, *The C++ Programming Language,* Addison-Wesley, 1986.

[37]   Thompson, Terrence, "Understanding Flat Cable - A Design and Assembly Guide," *Assembly Engineering,* July and August 1978.

[38]   *Unigraphics II,* McDonnell Douglas Manufacturing Industry Systems Company, 1991.

[39]   Wang, W. P., *Solid Geometric Modeling for Mold Design and Manufacture,* Technical Report 44, Cornell University, January 1984.

[40]   Winston, P. H., *Artificial Intelligence,* Addison-Wesley, Reading, Massachusetts, 1977.

Chapter 8
THE EXPERT COST AND MANUFACTURABILITY GUIDE: A CUSTOMIZABLE EXPERT SYSTEM

Phil London, Blair Hankins, Mark Sapossnek, and Steve Luby

ABSTRACT

Very few expert system applications have been distributed widely beyond the boundaries of the organizations within which they were developed. Instead, expert systems typically address problems for which local expertise dominates the methods for solutions to problems, and thus these expert systems represent idiosyncratic solutions that are applicable only within the organization that developed them. The obvious, though not easily attainable, solution to this problem is to provide to the end user of such an expert system the ability to customize its knowledge base. In this paper, we present the Expert Cost and Manufacturability Guide (ECMG) from this perspective. ECMG is an expert system designed to provide mechanical engineers with first-order manufacturing cost estimates and manufacturability feedback very early in the design process, during preliminary design. We describe the architecture of ECMG, particularly those aspects of its design that accommodate the need for customizability. This is followed by a description of the expert systems design methodology we employed to permit us to construct a customizable expert system application.

8.1. INTRODUCTION

By their very nature, expert systems primarily embody expertise that is quite specialized. A major problem in the commercialization or widespread acceptance of an expert system is that the expertise contained within the system rarely is valid beyond the organization that contributed the expertise. Although

Artificial Intelligence in Engineering Design
Volume III

223

problems for which expert systems provide attractive solutions cut across organizational boundaries, the specialized approaches and data used to solve a problem by an individual organization dominates the problem-solving techniques embodied in most expert systems.

Manufacturing cost estimation and manufacturability analysis is such a problem. Every manufacturing organization has its own idiosyncratic manufacturing processes. An evaluation of the cost of a product design or whether it is manufacturable relies on specialized and often proprietary information. Thus, an expert system to perform cost estimation or to provide manufacturability feedback must capture the essence of the cost estimation process in a generic way but must be customizable or extendable by the end user.

This paper describes the Expert Cost and Manufacturability Guide (ECMG) -- a generic expert system for cost estimation and manufacturability feedback. ECMG is intended to provide first-order estimates for mechanical designs early in the design process. ECMG also provides early feedback on manufacturability concerns such as process limits or design inconsistencies.

During conceptual or preliminary design, a large proportion of the ultimate manufacturing cost of a product is committed. The specific need that ECMG addresses (cost estimation for preliminary designs) arises from the fact that, at this stage of design, the engineer is primarily concerned with functionality and performance, and not with cost or manufacturability. While there are tools to assist the mechanical engineer with analysis of a design with respect to function and performance, no such tools exist in an accessible form for analysis of cost or manufacturability.

The Expert Cost and Manufacturability Guide is typical of expert systems that have broad applicability while requiring quite specialized knowledge to be useful to any particular end user. The challenge in designing ECMG involved:

1. identifying the generic aspects of the first-order estimation and manufacturability analysis problem,

2. "hardwiring" these into a generic expert system shell for cost estimation, and

3. providing tools to permit the customization of the generic shell into an expert system for cost estimation and manufacturability analysis specifically tailored to the user's environment.

To achieve these objectives for the Expert Cost and Manufacturability Guide, two distinct software packages were developed. The first (ECMG) is an "end user" application to be used by the practicing mechanical engineer during the preliminary design process. The second (ECMGEdit) is the generic expert system shell that permits a "lead user" to build specialized cost models to be utilized by ECMG when an engineer desires an estimate.

This paper is divided into two major sections. The first discusses the two software packages of the Expert Cost and Manufacturability Guide in detail. The second section discusses the problem of customizable expert systems in general, suggesting design guidelines for customizable generic shells.

8.2. THE EXPERT COST AND MANUFACTURABILITY GUIDE

The problem of cost estimation has traditionally been the province of manufacturing engineers. Typically, it requires substantial knowledge about the manufacturing processes, the specific machines, some understanding of the design function, and a capability to generate a detailed process plan or routing. Most of the techniques for cost estimation, both manual and semi-automated, assume this detailed approach to cost estimating. Of these methods, the greatest emphasis has been on developing models for estimating machining costs [6, 4].

There has been a consensus in the past few years that it is necessary to bring information regarding manufacturability and manufacturing cost estimates to the mechanical engineer. The objective of any such method or program must be to abstract from the basic manufacturing principles a set of geometric parameters and cost drivers that control cost and manufacturability. A comprehensive study conducted by Battelle [5] for the Air Force concentrated on isolating the key cost drivers for a suite of manufacturing technologies applicable to airframe manufacture. Research work by both Boothroyd [1] and Poli & Knight [7] led to the development of cost methods that rely on an abstract characterization of assembly and forging processes.

With respect to cost estimation, little has been attempted using expert system techniques. There has been an increasing volume of research work attempting to apply AI techniques to the problem of designing for manufacturability. In general, most of the effort has been to develop specific models that assess manufacturability for a single process. Luby's [3] casting design program and a system for designing extrusions [2] are examples of such focussed efforts.

The Expert Cost and Manufacturability Guide was designed to be a general, customizable tool, applicable to a variety of manufacturing processes and organizations. The Guide consists of two software packages, allowing it to achieve the desired level of customizability. The first package, ECMG, is used by a mechanical engineer to derive cost estimates of preliminary designs and to understand the manufacturability factors influencing the design. By using ECMG, the engineer can examine the effect of design trade-offs on cost and manufacturability. This is important because, while committing a large proportion of the ultimate manufacturing cost of a product, the engineer, during preliminary design, is primarily focussed on functionality and performance, not cost and manufacturability.

The second package, ECMGEdit, provides the tools necessary for "lead users" to make substantive changes and additions to the knowledge base, or "cost models", utilized by ECMG during an analysis session. ECMGEdit provides a knowledge base manager for maintaining the rules, relations, and equations (the language elements of the ECMGEdit language) stored in a cost model, and structured editors for the various elements of the ECMGEdit language. It is expected that ECMGEdit users will be the individuals who possess the specialized manufacturing expertise within an organization. ECMGEdit allows their expertise to be distributed to the design engineering community in the form of cost and manufacturability models.

8.2.1. ECMGEdit

ECMGEdit is essentially an editor for cost models. A cost model is a database of characteristics, equations, rules, relations, and restrictions. The cost model database is partitioned into contexts, which index the database according to a selected manufacturing process, material, or design feature. The representations for processes, materials, and features are hierarchical; for example, a cost model can represent the inheritance relationship (i.e. differences and similarities) between the generic manufacturing process "sheet metal forming" and its more specific sub-processes "stamping" and "fabrication."

The fundamental organizing entity within a cost model is a characteristic. Characteristics represent those cost drivers for a manufacturing process or material that determine the effects of design parameters on cost and manufacturability. Examples of characteristics include

---

```
"Minimum Wall Thickness" in Castings

"Dimensional Tolerance" in Stampings.
```

Of course, there are many others. In a cost model for a typical manufacturing process, one might expect to include 10-25 characteristics. These characteristics together define the aspects of a design that have first-order effects on cost. Equations (whose input format is very similar to FORTRAN arithmetic expressions) combine the values of characteristics supplied by the user (the mechanical engineer using ECMG to evaluate a design) to derive base cost estimates. The hierarchical inheritance relationships among processes, materials, and features permit characteristics and their associated equations to be placed in a context indexed by the most general process or material applicable. For example, many of the characteristics needed for an estimate of a die casting are valid, in general, for all castings and these characteristics and their equations can be stored in the casting context rather than being repeated for each of the sub-processes (e.g., die casting, investment casting, sand casting) of casting.

This organization of data also allows specializations of the definition of a process, material, or design feature according to the context. For example, the set of cost drivers and cost equations for a round hole can be dependent upon both the process and material selections. Therefore, the boolean characteristic: "Normal To Parting Plane?" would be part of a round hole description if the primary manufacturing process was either casting or forging, but would not be included for sheet metal forming processes. This prevents the system from requesting useless or irrelevant information about parting plane orientation when the mechanical engineer is costing a sheet metal design.

Equations are used not only to compute base cost estimates, but also to compute default values for characteristics. These defaults can be simple, or they can be arbitrarily complex. A default equation for the "dimensional tolerance" characteristic in stampings might be as simple as

```
DimensionalTolerance = .08
```

It may, on the other hand, account for complexities such as material thickness, material type, size of part, etc. The default computations provide a powerful mechanism for the ECMG model builder to generate recommended or economical settings for design parameters.

The equations in a model define base costs and default values for cost drivers that contribute to cost. The ECMGEdit user can also define rules to represent manufacturing constraints and policies that are not included in the general process model. These rules typically conclude costing data that is based on relationships among the individual process or feature models and modify the base costs generated by the models. The rules are indexed by the context and trigger opportunistically as information about the design is provided by the mechanical engineer.

The rules are capable of stating cost relationships that are more complex than can be handled by the default equation mechanism. For example, the following rule penalizes the cost of the tooling by 1.5% for every non-trivial hole in an aluminum forging (this rule is in the forging/aluminum context and therefore needs no screening clauses to test for that combination).

```
IF:      h = some(hole)
         diameter(h) > .125
THEN:    multiply(cost (primaryprocess), 1.015)
```

Rules may conclude numeric penalties or can assert one of the user-declared relations. These relations are usually created to hold significant intermediate results of a rule execution. An example of a relation among form features is "high tolerance." A rule may conclude a high tolerance relation between two holes which may drive another rule to conclude cost penalties to each hole. Like characteristics, equations, and rules, relations are indexed by context and can be inherited.

ECMGEdit provides a facility for the lead user to state process limits and thereby prevent the mechanical engineer from costing unmanufacturable parts. These restrictions provide a powerful facility to incorporate industry or company-specific design-for-manufacturability constraints in a cost model. Example restrictions are:

- You cannot use commercial stamping processes to achieve tolerances tighter than .01 times the sheet thickness.

- You cannot die cast ferrous metals.

When ECMG executes the model, triggering these restrictions will either eliminate process, material, or feature choices or will post warnings to the mechanical engineer stating the violated manufacturability condition. These restrictions are also indexed by context and can be inherited from parent processes, materials, and features.

8.2.2. ECMG

ECMG is the program that allows the mechanical engineer to estimate the cost of a component or assembly.

ECMG is driven by the cost model developed using ECMGEdit. The objective is to have the mechanical engineer provide a description of a component in sufficient detail to generate a first order cost estimate.

The cost models represent the significant cost drivers for a process as a set of characteristics and a set of cost computations based upon these characteristics. In a well-structured model, the set of cost drivers are carefully defined so that the engineer is required to describe the component and its features (that is, the design), and not the manufacturing processes. The cost guide employs a linearity assumption in generating an unordered, skeletal process plan for manufacturing a component and its design features. The cost contributions from each element of the plan are calculated using data in the cost model and are summed to produce total process, material, and tooling costs for the component.

The restrictions supplied in the cost model guide the mechanical engineer in the selection of the primary process, material, and design features. Then, by asserting values or by overriding default values for the cost drivers identified in ECMGEdit, the description of the component is completed and costs generated. During the detailing of the component, rules may activate to add costs, or restrictions may alert the mechanical engineer to questionable manufacturability conditions.

Key to the operation of ECMG is the use of a dependency network to maintain the equations, rule instances, and restrictions for a costing. As default equations are encountered in a context, they are instantiated into the dependency net to maintain consistency among the parameters upon which the equation depends. Rules that fire are also incorporated into the dependency net; this provides the truth maintenance facilities that allow the effects of rules to be retracted if the component description is modified. The dependency net also provides the base representation for the explanation system. This subsystem is capable of traversing the dependency net and generating explanations for the value of any parameter in the cost model.

The costing process creates dependencies between the geometry in the component and parameters in the cost model. Changes to the geometry or material characteristics are propagated into the cost model and the costs are modified or restrictions trigger to alert the engineer to manufacturability restrictions. By maintaining a complete dependency structure, the validity of the plan can be tested as changes are made to the geometry or component characteristics.

Maintaining this dependency information permits extensive exploration of design alternatives. The mechanical engineer can change any parameter, including the process, material, or feature selections, and observe the behavior of the cost model. Via the explanation mechanism, the engineer can interrogate the value of any cost driver or final cost and obtain information about the rules or equations involved in the derivation of that cost or value. The dependency management architecture provides a powerful "What-If" capability for the designer so that effects of design alternatives on cost of manufacturability can be rapidly assessed.

8.3. CUSTOMIZABILITY

When work was begun on the Expert Cost and Manufacturability Guide, the problem domain was examined to determine its appropriateness as an expert system application. There are several sources for compiled lists of problems characteristics that suggest appropriateness of a problem for an expert system approach (e.g., Prerau [8]). The application domain of first-order manufacturing cost estimation failed on several important measures of appropriateness:

- Is the task narrow and self-contained? Cost estimation is not a well-bounded domain; users will want to consider new classes of problems that a system was not designed to handle.

- Availability and uniqueness of expertise: Are there recognized, universally-acknowledged experts who are available to you? Would users or experts agree on whether the system's results are good? The actual and estimated manufacturing costs for a component vary substantially depending on location, machine availability, and company-specific processes; we have received tooling quotes for some sample parts that vary by factors of two and three.

- Is the expertise concentrated? That is, is the expertise obtainable primarily from a single expert and will it not require knowledge from a large number of areas? In fact, cost estimation expertise is very localized; a machining cost estimator usually understands little about injection molding processes.

For the problem addressed by the Expert Cost and Manufacturability Guide, these considerations suggested that formidable barriers would be encountered when trying to develop a turn-key expert system for manufacturing cost estimation. An obvious solution was to permit the user to customize or extend the knowledge base so the application could adopt local costing data and methods for incorporation into its operation. The key issue was to derive a design that could strike a balance between a body of built-in costing capabilities and the ability for a user to make significant alterations to the knowledge being brought to bear to solve a costing problem.

To design a customizable, generic cost estimation expert system, we began with a set of requirements for customizability:

1. Customization and extension of the knowledge base should be performed in a language that is natural to the user.

2. The knowledge base should be decomposable so as to reduce complexity.

3. Generally accepted (universally utilized) concepts and methods from the application domain should be "hardwired" into the application.

4. Limit flexibility and power available to the user to the extent that the user's task is achievable.

The Expert Cost and Manufacturability Guide (specifically, ECMGEdit) successfully achieves each of these objectives, and we believe the result is a system that achieves its overall objective -- customizability by the end user.

The first customizability objective, customization in a language natural to the user, was achieved by confining the format used to encode knowledge for the bulk of the information from the costing domain to algebraic equations. Thus, most of the information specified is in a straightforward format quite familiar to any engineer. Predicates, such as restrictions or the premises of rules are expressed as straightforward boolean expressions; these require more sophistication on the part of the user, but are well within the reach of any individual who has even minimal computer programming experience or training in formal logic. Our preliminary observations on the use of ECMGEdit suggests that by avoiding a knowledge base language that admits complex procedural semantics (as one might find in a conventional expert system shell), we have greatly reduced the difficulty of the user's task of becoming proficient in the language.

The second customizability objective addresses the issue of reducing the complexity of managing the knowledge base. This is achieved by partitioning the knowledge base into contexts that are meaningful to the user. In ECMGEdit, these contexts are indexed by a selected manufacturing process, material, and design features. One obvious advantage of this scheme is to eliminate the need for screening clauses (i.e., clauses in rule premise expressions whose sole purpose is to limit the applicability of a particular rule to a particular computational context). In fact, by eliminating screening clauses such as:

```
IF:      PrimaryProcess=="DieCasting"              Material=="Brass",
```

from all rules within a context, equations serve the same role as rules within a less structured system; equations are evaluated only when the current computational state places the system within the context for which the equation is defined. Thus, the effect of simple statements in the language are enhanced by the additional semantics imposed by the context decomposition. Contexts also impose a natural organizing principle on the knowledge base, which suggests particular problem-solving architectures to the user.

The third objective, that generally accepted concepts and methods from the domain be hard-wired into the application, distinguishes ECMGEdit from a general-purpose expert system shell. ECMGEdit is a shell for building manufacturing cost estimation expert systems, only. This is true because the system "knows," or has built-in semantics for, concepts such as process, material, feature, and process limit. It also has built-in methods to decompose costs into tooling, material, and labor and can compile a bill of materials for a set of components. The rule language has been optimized to best represent the knowledge employed by experienced cost estimators.

The final customizability requirement is to limit the power available to a user of the system. This may seem counter-productive, but we have made the assumption that a typical user of the system is relatively inexperienced as a computer programmer, and totally inexperienced with respect to expert systems. By limiting flexibility and power, a greater commitment to a particular problem-solving architecture is possible in the base tool. This, in turn, requires less design to be required of the user for problem-solving methods and architecture. An example of a limitation in the power of the ECMGEdit language is the capabilities of the conclusions of rules. In ECMGEdit, rules are permitted only to make multiplicative or additive modifications to cost estimates, or to make simple relational assertions. These limitations preclude the ability to do meta-planning or to generate arbitrary side-effects. The user of ECMGEdit is focussed on the problem of describing the knowledge required to generate cost estimates; the user is not distracted by issues of problem solver design or architecture.

8.4. SUMMARY

The Expert Cost and Manufacturability Guide is being developed as an adjunct to Cognition's Mechanical Advantage system [9]. Mechanical Advantage (MA) is a mechanical design (MCAE) workstation that supports preliminary design by permitting freehand sketching, automatic recognition of geometric constraints, and definition of an engineering model via an integrated equation solver that defines mutually constraining information to the geometric modeler. The Expert Cost and Manufacturability Guide shares the MA's user interface and can extract geometric information from MA sketches (though it does not do feature recognition from the geometry).

8.5. ACKNOWLEDGEMENTS

This work was done when the authors were with Cognition, Inc., 900 Tech Park Dr., Billerica, MA 01821.

8.6. REFERENCES

[1]   Boothroyd, G and Dewhurst, P., "Design for Assembly - A Designer's Handbook," Department of Mechanical Engineering, University of Massachusetts, Amherst, 1983.

[2]   Libardi, E.C., Dixon, J. R. and Simmons, M. K., "Designing with Features, Design and Analysis of Extrusions as an Example," *Proceedings of the Mechanical Design Conference*, Chicago, March 1986.

[3]   Luby, S.C., Dixon, J. R. and Simmons, M. K., "Designing with Features," *Computers in Mechanical Engineering,* Vol. 5, No. 3, November 1986.

[4]   Mathews, L. M., *Estimating Manufacturing Costs,* McGraw-Hill, New York, 1983.

[5]   Noton, B. R., *ICAM Manufacturing Cost/Design Guide,* AFWAL-TR-80-4115, Air Force Wright Patterson Labs., September 1980.

[6]   Ostwald, P. F., *American Machinist Cost Estimator,* McGraw-Hill, New York, 1985.

[7]   Poli, C. and Knight, W. A., "Design for Forging," Department of Mechanical Engineering, University of Massachusetts- Amherst, 1985.

[8]   Prerau, D. S., "Selection of an Appropriate Domain," *AI Magazine,* Vol. 7, No. 2, Summer 1985.

[9]   Villers, P., "Computers for Conceptual Design," *Computer-Aided Engineering (USA),* May 1986.

Chapter 9
ENGINEOUS: A UNIFIED METHOD FOR DESIGN AUTOMATION, OPTIMIZATION, AND INTEGRATION

Siu Shing Tong, David Powell, and Danny Cornett

ABSTRACT

Engineous, the generic software shell described in this chapter, provides a new method for unifying the automation, optimization, and multidisciplinary integration of product design. Engineous combines advanced computational techniques -- such as machine learning, expert systems, and object-oriented programming -- with such conventional techniques as numerical optimization. Using captured human knowledge, Engineous automatically invokes the appropriate analysis codes and design procedures, deduces the key design parameters, and manages the use of optimization packages. When needed, genetic algorithms are used to uncover unconventional designs.

Engineous was tested on a set of six engineering optimization problems that have been proven difficult to numerical optimization techniques. Results demonstrated that it was more robust and efficient than either genetic algorithms or numerical optimization used in isolation.

Engineous has produced designs with higher predicted performance gains than current manual design processes, has reduced turnaround time by 10-to-1 (on average), and has yielded new insights into product design. It has been used in a wide variety of applications, including the design of aircraft engine turbines, molecular electronic structures, cooling fans, DC motors, and electrical power supplies, and the concurrent preliminary design, aerodynamic detailed design, and mechanical detailed design of 3D turbine blades.

Artificial Intelligence in Engineering Design
Volume III

235

9.1. INTRODUCTION

As product design and the tools used in product design become more complex, software tools to make the design process more efficient become increasingly important. Although many of the new, complex computational codes are used routinely for new product design, their use is less efficient than it should be. Engineers who are knowledgeable about product design are rarely skilled in the application of these new codes. The few engineers who are skilled in both disciplines are faced with the tedious and time-consuming process of manually iterating analysis codes to obtain optimum design. For most complex engineering products, there is an additional difficulty. A large number of engineers with diverse expertise have to work together and communicate diverse kinds of information.

The Engineous software shell described here addresses all of these problems. It uses the knowledge of the experienced design engineer, invokes the appropriate analysis codes, automates the iteration of the codes, and provides a platform for exchanging information from diverse disciplines.

9.2. PRODUCT DEVELOPMENT CYCLE

Typically product development consists of a cycle of Conceptual Design, Preliminary Design and Detailed Design. Although these stages may have different names or may be combined differently in different organizations, the following sequence is typical.

1. *Conceptual Design:* Uses some quick "hand" calculation to select among the major alternatives the most promising approach/configuration to meet the requirements of the product. Innovative concepts are often generated here. Conceptual Design takes on the order of 1% of the total design effort.

2. *Preliminary Design:* Uses simplified (1D, 2D, partial physics) analyses to evaluate trade-offs between components and between disciplines. Preliminary Design defines each component's goals, requirements, and certain key parameters. Preliminary Design takes on the order of 5% of the total design effort.

3. *Detailed Design:* Uses accurate analysis codes to come up with a manufacturing drawing for making a particular part that meets the

Preliminary Design requirements. Detailed Design often takes at least 90% of the total design effort.

Conceptual design involves abstract concepts and innovation where automation or optimization may be impractical and inappropriate at present. But for complex systems, a tool flexible enough to assist in the examination of various concepts is valuable. Preliminary design involves a large number of parameters, multidisciplinary knowledge, and examination of many scenarios. Preliminary design can benefit greatly from a tool that can automate some tedious iteration processes; however, non-analytical information is often required to automate this process.

In the later design stages, goals and requirements are often well defined in mathematical terms, and mathematical tools [5, 10] have been developed to optimize individual parts or manufacturing processes. However, problems remain in applying those optimization tools to a large number of continually evolving analysis codes and in information exchanges between various disciplines to ensure consistency.

The absence of integration between these design stages presents a major obstacle in rapid and efficient product development. One generally recognized problem is inattention to manufacturing considerations during design stages, which results in the selection of designs that are difficult to manufacture. In conceptual design and most preliminary design, only a small fraction of the final product is defined and predicting manufacturability may be difficult. In detailed design, there is limited flexibility in making changes that may affect other components. A significant detailed design change in one component often requires starting the preliminary design process all over again. There is little question that design productivity and quality could be improved significantly by integrating, automating, and optimizing the design processes both vertically, across design stages, and horizontally, across disciplines.

9.3. ENGINEOUS SHELL

The Engineous project at GE Corporate Research and Development was undertaken to develop a new approach to product design. Engineous is a generic shell that combines expert systems, numerical optimization, and machine learning technologies (for function optimization) to provide a unified method for using computational analysis codes in the automation, optimization, and integration of product design. The basic concepts of Engineous may be summarized as follows:

- *Unified automation, optimization, and integration:* A single system has the capability to automate the design process by emulating designer engineers, the capability to explore the design space without knowledge, and the capability to apply both automation and optimization at a more global level, integrating a number of disciplines and analysis codes.

- *Generic shell:* Problem-specific information is separated from generic design functions and is organized as knowledge bases.

- *Combination of expert system, numerical optimization, and machine learning technologies:* A hybrid approach is provided to accommodate a wide range of engineering problems having various combinations of continuous, discrete, and symbolic parameters, various amounts of design knowledge, and drastic differences in the shape of the objective functions.

- *Rapid and non-intrusive coupling to computational analysis codes:* The rapid increase of computer power has resulted in a proliferation of new or modified computer-aided analysis (CAE) codes. Engineous's ability to run these codes without modification is essential in today's engineering environment.

A detailed description of the design automation, optimization, and integration approaches used by Engineous follows.

9.3.1. Design Automation

Generic functionality is provided for the following capabilities, which are required to automate CAE-based design:

- Preparing input files
- Managing the execution sequence and data flow between programs
- Extracting relevant output parameters
- Applying captured design procedures to achieve design objectives

Problem-specific information for each application is captured in a Design Knowledge Base within Engineous. Each application will have its own Design Knowledge Base, which can be further decomposed logically into: Design Parameter KB, Program KB, Program Sequence KB, and Design Rules KB. Figure 9-1, a diagram of the Engineous Design Knowledge Base, shows all of these generic modules and design modules, rules, and parameters and program sequence for a particular application.

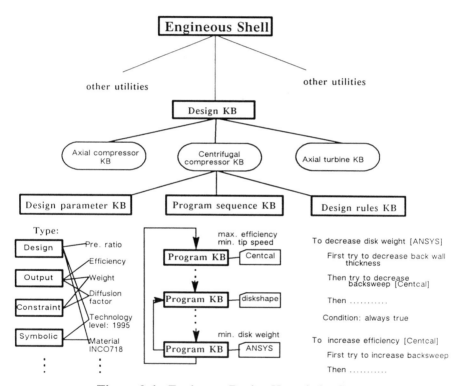

Figure 9-1: Engineous Design Knowledge Base

(Names of generic modules are in bold type in rectangular boxes; the three design modules, the design rules, design parameters, and program sequence, shown in light type, are examples from a particular application.)

9.3.1.1. Design parameter KB

In this KB, each parameter is classified as belonging to one or more of the 12 default object types (e.g., design, output, constraint) with various attributes. For example, a "design" parameter has current value, minimum increment, default value, etc. The "material" design parameter is also a "symbolic" parameter that can only take on a predefined set of symbolic values. The tip-speed, rpm, and tip-diameter parameters are coupled parameters (i.e., *tip–speed=constant×tip–diameter×rpm*), whose relationship is stored; any change in one parameter affects the others.

9.3.1.2. Program KB

This KB captures information on: how to set up the input files, how to execute an analysis code, what the code's input and output parameters are, and how to extract useful information from analysis output files for Engineous. The Program KB also manages all pre- and post-processing programs that need to be executed for each analysis code.

9.3.1.3. Program sequence KB

This KB manages the execution sequence. It does not contain a hardwired procedure; it contains only the allowable transitions between programs; i.e., for each program, what programs besides itself can follow it. Which program actually runs is determined at run time depending on the parameters to be modified.

9.3.1.4. Design rules KB

This KB captures human design modification knowledge in the form of logical steps a designer would go through. Engineous uses a special rule syntax developed to represent analysis-code-based iterative design knowledge. This knowledge representation is based on the observation that only a limited set of goals, premises, and actions are used during iterative design. An Engineous design rule might be paraphrased as follows:

```
Goal:          To increase X
Conditions:    Y > 3 and
               Z is not at its upper limit
```

```
Then try the following actions in order:

Action 1. Separable:    up Z by 10%,
                        down A by 2,
                        set Material to "inco718",
                        vary B, C, and D

Action 2. Inseparable:  set Z to 50,
                        set Material to "tin125",
                        up T,
                        vary C

Action 3. Sequential:   set C to 5.5,
                        vary D
          Etc. ...

This rule is weighted at: 90
```

where X is an output parameter that may be a part of the optimization function (e.g., turbine efficiency) or may have lower bounds (e.g., minimum flow angle), and Y and Z in the Conditions part may be input or output parameters.

The actions of the rule will be modified if Engineous determines that the current goal is to increase X and the conditions specified are not violated, the weight of this rule is higher than other rules that can be activated, and the rule is not currently suspended because of previous repeated failures.

The action-type "separable" tells Engineous to vary one or more parameters within the action even if some others are not allowed to vary for this particular run. For Action 1, Engineous will increase Z by 10% of its current value if it has no upper bound, or 10% of the difference between its current value and its upper limit. It will subtract 2 from A and set Material to "inco718." Then it will pass B, C, and D as variable parameters to the optimization module in order to maximize X. If this action is not successful (i.e., the value of X is not increased), the action is suspended and Engineous will restore the values of Z, A, Material, B, C, and D and go to Action 2. No part of Action 2, which is an "inseparable" action, will be applied if any one of the parameters is not a design variable or any one of them is already at its boundary. "Sequential" actions are to be taken one at a time. If an action is successful, Engineous will repeat it until it reaches the point of diminishing returns.

9.3.2. Design Optimization

The Engineous optimization module tightly couples three technologies -- expert systems, numerical optimization, and machine learning -- switching from one to another at various stages of a single design run. Individually, each approach has advantages and disadvantages. In combination, they complement each other in many ways.

9.3.2.1. Expert systems

The Design Rules KB can not only capture procedural design knowledge that must be emulated to automate the design process, but can also capture design knowledge that can make Engineous more efficient. If the latter form of knowledge exists, which often comes in the form of which subset of parameters to try first given certain conditions, Engineous will first draw on it to emulate design experts. This use of an expert system yields the greatest gain with the smallest number of analyses and produces designs that engineers have confidence in. Expert systems can also explain how and why a final design was chosen, which is important for balancing predicted performance against other factors not modeled in the analysis codes. Since the human design knowledge is often incomplete or biased, however, a pure expert system is very likely to miss better designs. Moreover, a pure expert system for complex design problems can be expensive to build and maintain. Therefore, Engineous supplements incomplete knowledge with some heuristic search techniques to emulate the human trial-and-error process. These techniques are invoked when human design knowledge is exhausted.

9.3.2.2. Numerical optimization

A numerical optimization package called ADS (Automated Design Synthesis) [11, 12] can be invoked automatically when the expert system and trial-and-error process cannot make any more gains. Engineous can also cycle through a series of optimization plans, with each plan containing different ADS options. The advantage of numerical optimization is that it can be applied to a wide range of problems. ADS has proven to be effective for reaching a local optimum for a smooth, continuous objective function. For problems with a large number of parameters and CPU-intensive codes, however, the large number of gradient calculations required can be prohibitively expensive. ADS is sensitive to the initial guess and internal parameter settings. Having expert systems to identify key parameters for various situations, and having Engineous cycle through a sequence of ADS runs with various ADS options, has partially eliminated these disadvantages.

9.3.2.3. Machine learning

The use of machine learning methodologies for function optimization was suggested by [1] and [4], and successful demonstrations of such an approach to turbomachinery design have been shown [9]. Engineous employs the "genetic algorithm" machine learning methodology [2, 3]. The basic idea of the genetic algorithm is not to make absolute decisions as to whether a trial design should be kept or discarded, but to penalize or reward trial designs on the basis of the value of the optimization function. The process starts with a set of initial designs, called a generation, which Engineous seeds with some good designs generated by expert systems and numerical optimization. New generations of the same number of trial designs (say, 50) are generated first by "selection," then by "crossover," and finally by "mutation." During a selection process, designs that have a higher optimization function value have a better chance to survive and reproduce themselves, while designs with a lower value are more likely to be eliminated. Crossover exchanges a random set of parameters of some past trial designs with other designs to generate new designs. This step is constructed in such a way as to make good features tend to cross over together. A random mutation of a few trial designs takes place periodically. The mutation process is implemented by flipping bits with a binary coded representation of the design vector. After a sufficient number of generations, clusters of good design regions -- occasionally with innovative answers -- will evolve.

Because the genetic algorithms are not gradient based, they will not be trapped in local optima or constraint boundaries and can handle a wide range of parameter types, including symbolic parameters (e.g., material, technology levels). On the other hand, genetic algorithms do not explore local optima and require many orders of magnitude more analysis runs to obtain useful results. Genetic algorithms are, however, very amenable to parallel or distributed processing.

9.3.2.4. Interdigitation of expert systems, numerical optimization, and genetic algorithms

Expert systems, numerical optimization, and machine learning are tightly coupled in Engineous and work together for a single design process. Unless modified by users, the Engineous design optimization module runs in three cycles. In the first cycle, only an expert system is used to get a good answer with a small number of runs. In the second cycle, Engineous alternates between an expert system and numerical optimization to perform local hill climbing until no further gain can be made. At that point, the solution may be at a local optimum, having been trapped in a complex constraint boundary or stopped due to premature convergence. The third cycle then invokes a genetic algorithm for a specified number of generations and, if a promising solution is found, switches back to the expert system and numerical optimization for local hill climbing. For a detailed description of the interdigitation of genetic algorithm and expert system technologies, see [6].

In a real-world design process, there is seldom enough time to explore all possibilities. The sequence of Engineous cycles ranks the parameters and search methods according to their expected gain per analysis run to allow maximum gain for a specified number of analyses run. A user can specify the total elapsed time, and Engineous will terminate and present the best results obtained thus far. Elapsed time is better than other arbitrary "convergence" criteria. Because most optimization processes are sensitive to the initial guess and search procedure, various starting points and search procedures should be tried until time runs out.

9.3.3. Design Integration

In order to perform design integration, a number of capabilities are required, for example:

- Managing the relationship between parameters that are the interface between components or disciplines (e.g., the rpm of two components sharing the same shaft should always be the same)

- Allowing incremental building of a large application

- Facilitating the trade-off between components and between disciplines

These capabilities are provided in Engineous through the use of object-oriented programming and a simple constraint propagation mechanism. All entities in Engineous -- including parameters, programs, design rules, search methods, etc. -- are represented by objects. Inheritance provides an efficient way to manage the complexity of a large-scale application. Simple constraint propagation is represented as methods activated by any change in parameters that are related to other parameters. A generic parameter-studies object is provided to study trade-offs. It can be used: (i) simply to vary some predefined parameters in a certain pattern, or (ii) to combine predefined patterns recursively to form a more complex study matrix, or (iii) at each complex matrix entry to re-do the whole design optimization sequence with a new set of variable parameters. Some designers view this parameter-studies feature as very useful.

9.4. HISTORICAL TEST CASES

A test set of six engineering optimization problems was used to demonstrate that Engineous's interdigitation approach is more efficient and robust than either genetic algorithms or numerical optimization used in isolation. The six engineering problems were part of a test set of 30 problems originally selected by [8] to analyze the performance of 25 numerical optimization codes. The numerical optimization codes performed poorly on six test problems because of discontinuities, multimodality, gradient insensitivities, equality constraints and scaling problems. Since no optimization code could solve more than two of these problems, Sandgren dropped the six engineering problems from his final comparative analysis. These six problems were selected for our analysis because they are a difficult test set, have known optima and are representative of the types of parameter space problems that are evident in real-world engineering problems (e.g, discontinuities, multimodality). The characteristics of the test problems are summarized in Table 9-1.

Engineous proved more robust and efficient than either numerical optimization using sequential quadratic programming, numerical optimization using the modified method of feasible directions or genetic algorithms used in isolation. (A detailed discussion of the performance of each technique on each problem is given in [7].) The number of test problems solved within 5000 runs of the simulation code by the best numerical optimization technique, genetic algorithms, and Engineous are shown in Table 9-2. A 5000-run limit was established to simulate the limited amount of time available for an engineer to obtain a design (e.g., a code which requires 1 minute to run can be optimized in 3 days). Engineous clearly outperformed numerical optimization and genetic algorithms. For example, for a total relative error of .5, Engineous solved five problems compared to one for numerical optimization and one for genetic algorithms.

9.5. ENGINEOUS APPLICATIONS

Back-to-back comparisons of designs using manual processes and designs using Engineous are available for a number of commercial product applications. The examples included here indicate the levels of complexity Engineous is currently dealing with as well as the impact it has had on design productivity.

Table 9-1: Characteristics of the Engineering Test Set

Problem	N	J	K	Feasible Starting Point	ASC	AOC	Characteristics
1 Chemical Reactor Design Sandgren #9	3	9	0	Yes	0	0	
2. Gear Ratio Selection Sandgren #13	5	4	0	Yes	0	0	Discontinuous objective function
3. Three-Stage Membrane Separation Sandgren #21	13	13	0	No	3	11	Small feasible region
4. Five-Stage Membrane Separation Sandgren #22	16	19	0	No	7	16	Small feasible region Poor scaling
5. Lathe Sandgren #29	10	14	1	Yes	4	3	
6. Waste Water Treatment Sandgren #30	19	1	11	No	4	12	Large number of nonlinear equality constraints

N – Design variables   ASC–Active Side Constraints
J – Inequality constraints   AOC–Active Output Constraints
K – Equality constraints

Adapted by permission from Davis, *Handbook of Genetic Algorithms*, p. 325 (Van Nostrand Reinhold: 1991).

Table 9-2: Number of Historical Test Problems Solved for Each
Optimization Approach for Different Relative Errors

Relative Error	Optimization Approach		
	I	NO	GA
.5	5	1	1
.25	4	1	1
.1	3	1	1
.075	2	1	1
.05	1	0	1
.01	0	0	1

Relative error $= \left| \dfrac{f(x) - f(x^*)}{f(x^*)} \right|$

where x^* is optimal design

I – Interdigitation
NO – Numerical Optimization
GA – Genetic Algorithm

9.5.1. Aircraft Engine Turbine Preliminary Design

A modern high-bypass engine turbine consists of multiple stages of stationary
and rotating blade rows inside a cylindrical duct. A typical large transport engine turbine has 7 stages and over 700 parameters in the preliminary design
phase; 100 of the input parameters are varied. There are a few dozen design
rules in this application. For example,

```
To increase blade exit flow angle:
    First try to decrease reaction,
    Then try to increase rpm,
    Then try to decrease loading.
```

This rule always applies.

This application was implemented while Engineous was being developed. The time it takes to install this application using the current version of Engineous is estimated to be around one to two man-months.

A multi-stage low pressure turbine was selected as a good candidate for testing the applicability of Engineous in a real-world design environment with a real design project. For the particular turbine in question, a design optimization procedure had already been started in which a designer was optimizing the turbine with a goal of 0.75% efficiency improvement. The Engineous design was started in tandem with that work. The designer achieved a 0.5% improvement in design in 10 man-weeks while the Engineous design achieved a 0.92% improvement in one week.

The power of the interdigitation approach is also shown with this turbine design application. Figure 9-2 shows the partial result of a thorough study for a new 2-stage turbine design. Over 30 ADS options with different search methods, gradient deltas, convergence criteria, and normalization were tested. The ADS work was stopped when no gain even as small as 10E-06 could be made. It can be seen that some ADS runs did much better than the expert system in this case. ADS performance was found to be highly dependent on the choice of ADS parameters, whereas an expert system always produces "good" results without tuning. Full interdigitation of all three search methods outperforms each of the three search methods by an efficiency gain of as much as 1%, a very significant number for turbine efficiency. The final turbine was an unconventional design with some of the parameter distributions opposite to what was done traditionally. Analysis of the optimization history shows that although the use of the genetic algorithm resulted in only a small gain in efficiency, it does appear to have pushed the optimization process away from being trapped in constraint boundaries so that the local hill climbing process could continue. Note that without the genetic algorithm, partial interdigitation -- expert system and ADS -- produced a lower-performance (0.5% less than that of full interdigitation) but conventional turbine.

9.5.2. Molecular Electronic Structure Design

This design task was to locate the lowest energy state of a molecular electronic structure. The design has fewer than a dozen variable parameters. The analysis code is relatively slow and has to be executed on a remote mini-supercomputer.

It took one man-week of effort to solve a simplified case by hand iteration. It took half a day to couple the analysis code to Engineous, and one day to produce a similar solution.

Optimization techniques

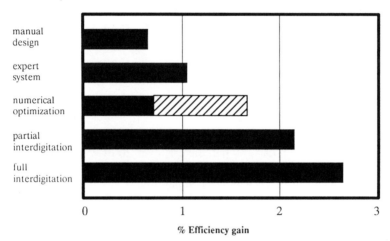

Figure 9-2: Efficiency Gain vs. Iteration for Various Search Methods

Expert system, ADS (numerical optimization; the shaded region representing range of performance depends on ADS parameter settings), interdigitation of expert system and ADS, and interdigitation of expert system, ADS, and Genetic Algorithm.

9.5.3. Cooling Fan Design

This is a very simple design task for Engineous, with only 44 parameters, 18 of which are input. The significance of this application is that the same problem has been solved with a pure OPS5 rule-based system, EXFAN [9]. The development of EXFAN from scratch took approximately two man-months. Reimplementing the same fan design with Engineous took half a day.

The design task is to determine the geometry, rotational speed, and air flow characteristics of a simple cooling fan. The objective is to minimize power consumption while maintaining a certain airflow. The quasi-3D fan analysis code uses 2D airfoil experimental data and requires aerodynamic knowledge to interpret the results. Ensuring a valid solution is the key issue. The designs generated by Engineous are slightly better than those developed by design experts. The turnaround time is less than 10 minutes for Engineous and half a day for the human expert.

9.5.4. DC Motor Design

The design task is to determine a complete specification for an industrial large DC motor. This task has 180 parameters, over 70 of which are input. This application took a few man-days to implement. The turnaround time of Engineous for this application is a few hours compared to a few man-weeks by the design expert.

9.5.5. Power Supply Design

Engineous has been coupled to HSPICE for testing its applicability to power supply design. A simple forward converter with a target voltage of 75 volts was designed both by a design expert and by Engineous. Engineous completed the design in 10 hours vs approximately 3 weeks and obtained a 75.08 volts design vs 71.7 volts. It took a few days to couple Engineous to this test problem.

9.5.6. Concurrent Preliminary Design, Aerodynamic Detailed Design, and Mechanical Detailed Design of 3D Turbine Blades

This application illustrates the type of complex multidisciplinary design task Engineous was developed to solve. The process calls for completing a preliminary design of an aircraft engine turbine, then obtaining a complete detailed geometry of all the turbine blade rows that meets the aerodynamic requirements of the turbine as well as mechanical vibration and static stress constraints. This application involves more than two dozen CAE codes, where some of the three-dimensional CAE codes, such as ANSYS and CAFD, are so complex that they may require a few man-weeks just to analyze one design scenario. A typical design cycle time, depending on the number of stages and complexity, is 12 to 24 man-months. A completely automated design under Engineous is expected to take a few weeks. The CAE code flowchart is shown in Figure 9-3.

9.6. SUMMARY

The current version of Engineous has demonstrated the profound impact such a system can have on productivity and performance. Engineous is emerging as a standard for turbine preliminary design at GE. It is also being used to design DC motors, electrical circuits, space power generators, and ground-based power generation plants. The aircraft engine turbine design application described here demonstrates the powerful concept of interdigitation of search strategies. Interdigitation allows Engineous to solve a wide range of problems, from well-understood tasks -- where there is ample knowledge -- to new design problems; from smooth to rough objective functions; from aiming for conservative design to exploring new concepts; and from problems with only real parameters to complex problems with a mixture of real, integer, and symbolic parameters.

Although using the genetic algorithm with other search techniques resulted in large gains in efficiency in a number of applications, its role is still not fully understood. More work is needed to understand the process and possibly extract a more efficient mechanism from the genetic algorithm.

With the rapid increase of computational power and the advance of numerical simulation methods, the number and complexity of engineering products that can be analyzed accurately will be increased rapidly and analysis turnaround time reduced substantially. Human intervention in computation-based designs has become a productivity bottleneck, preventing many powerful numerical simulation modules from being used to their full potential. In eliminating this

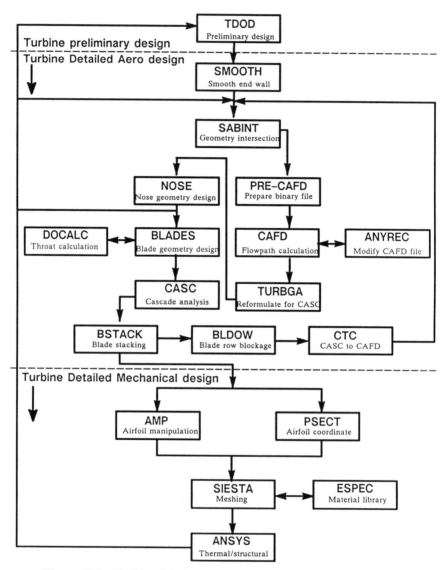

Figure 9-3: Turbine Blade Preliminary and Detailed Design with
Concurrent Aerodynamic and Mechanical Analyses

bottleneck, a system like Engineous will not only make the design process more productive but also, by relieving designers of the need to carry out tedious, iterative procedures, make it more fun.

9.7. ACKNOWLEDGMENTS

We want to thank Mr. Brent Gregory, manager of Turbine Aero and Cooling Technologies at GE Aircraft Engine, who contributed greatly to the success of the turbine preliminary design project; and Dr. Carol Russo, Dr. Robert Phillips, and Mr. Eugene Libardi, who provided technical support for other GE Aircraft Engine applications at Lynn. Professor Garret Vanderplaats has provided his ADS expertise. Engineous technology development is funded by the GE Research and Development Center. The turbine blade concurrent design problem is funded by DARPA.

9.8. REFERENCES

[1]   De Jong, K., *An Analysis of the Behavior of a Class of Genetic Adaptive Algorithms,* unpublished Ph.D. Dissertation, Dept. of Computer Science, University of Michigan, 1975.

[2]   Goldberg, D. E., *Genetic Algorithms in Search, Optimization, and Machine Learning,* Addison-Wesley, Reading, MA , 1989.

[3]   Grefenstette, J., "Genetic Algorithms and their Applications," 2nd International Conference on Genetic Algorithms, 1987.

[4]   Holland, J., *Adaptation in Natural and Artificial Systems,* University of Michigan Press, Ann Arbor, Michigan, 1975.

[5]   Kirkpatrick, C., Gelatt, C., and Vecchi, M., "Optimization by Simulated Annealing," *Science,* Vol. 220, pp. 671-680, May 13 1981.

[6]   Powell, D.J., Tong, S.S., and Skolnick, M.M., "Engineous, Domain Independent, Machine Learning for Design Optimization," *Proceedings of the Third International Conference on Genetic Algorithms,* Held at George Mason University, pp. 151-159, June 4-7 1989.

[7]   Powell, D. J., *Inter-GEN: A Hybrid Approach to Engineering Design Optimization*, unpublished Ph.D. Dissertation, Rensselaer Polytechnic Institute, 1990.

[8]   Sandgren, E., *The Utility of Nonlinear Programming Algorithms*, unpublished Ph.D. Dissertation, Purdue University, 1977.

[9]   Tong, S. S., "Coupling Artificial Intelligence and Numerical Computation for Engineering Design," *AIAA 24th Aerospace Sciences Meeting*, Reno, Nevada, January 6-9 1986.

[10]   Vanderplaats, G.N., *Numerical Optimization Techniques for Engineering Design: With Applications*, McGraw-Hill, New York, 1984.

[11]   Vanderplaats, G. N., "Lecture Notes on Computer Aided Optimization in Engineering Design," Union College Continuing Education Course, July 27-31, 1987.

[12]   Vanderplaats, G. N., "ADS -- A Fortran Program for Automated Design Synthesis," Engineering Design Optimization, Inc., Santa Barbara, 1988.

Chapter 10
A UNIFIED ARCHITECTURE FOR DESIGN AND MANUFACTURING INTEGRATION

Steven H. Kim

ABSTRACT

This chapter discusses the functional requirements and constraints for an effective production environment, ranging from design to manufacturing. Generic approaches to modularity are presented, followed by a two-dimensional architecture to support computer-integrated manufacturing. This architecture serves as a skeleton on which to base conventional information structures as well as artificial intelligence techniques. The utility of this approach is demonstrated by a number of examples that show how the structure can accommodate specific facets of design and production.

10.1. INTRODUCTION

The roots of industrial productivity are found in the application of novel technologies to the production environment. If we are to fully harness new information technologies to design the factory of the future, it would be insufficient to introduce the new technologies in a haphazard fashion.

To ensure the effective utilization of the new technologies, we must develop conceptual frameworks [5, 6, 8, 11], mathematical models [1, 3, 6, 7] and organizational infrastructures [9, 10, 13] to accommodate the technologies so that they may work synergistically to produce an effective production system. This chapter presents an organizational architecture to allow for the fusion of the existing and emerging technologies. More specifically, it describes a multimodal nested architecture consisting of hierarchical and layered structures. This ar-

Copyright © 1992 by Academic Press Inc.
All rights of reproduction in any form reserved.
ISBN 0-12-660563-7

chitecture accommodates three major subsystems: system software, system hardware, and production materials.

10.2. REQUIREMENTS FOR A FACTORY

The high-level functional requirement of a factory is to produce goods to specifications. The constraints on the factory are as follows:

1. Capital investment requirements should be competitive.
2. Production costs should be competitive.

The current turbulence in the industrial sector, as well as the attendant uncertainty for managerial and engineering decision-making, stem from the following developments:

1. Consumers are demanding higher quality (reliability, availability and simplicity of maintenance). This requires redesign of existing products and enhanced quality control in production.
2. Consumers expect a wider spectrum of products better tailored to individual needs. This implies the ability to produce goods in increasingly smaller lot sizes at competitive cost.
3. Competitive firms are introducing new products, or improvements in old products, at an accelerating rate. The decrease in product lifecycles implies the need to shorten response times in both the design phase and the production phase. Moreover, the increase in product diversity introduces the need to employ highly flexible capital equipment.
4. Competitors are making full use of the learning curve in the production phase to lower both direct and indirect costs to increase the margin on revenues. This is often achieved by the redesign of products and processes, by the replacement of human labor, by enhanced production planning and control, and by better utilization of existing information through computer integration.

To address these requirements, the modern factory must have these characteristics:

1. Improved integration of design and production phases.
2. Tools to enable the rapid design or redesign of products and processes.
3. Flexibility in capital equipment, such as intelligent machine tools and robots.

"Improved integration of design and production" refers in part to the incorporation of manufacturing knowledge in the design phase to enhance design for producibility. Another aspect of system integration relates to the pipelining of operations: for example, dies for certain components may be prepared even before other aspects of the overall design are fully finalized.

The way to attain a more flexible, adaptive factory that can respond quickly to the changes in the environment is to build on an effective architecture. This architecture, in turn, should fulfill the following functional requirements:

1. Serve as a unifying framework that accommodates hardware and software, including a range of fabrication technologies.
2. Exhibit a modular organization. This allows for the containment of complexity and ease of reconfiguration in adaptating to changing requirements.
3. Provide an open architecture to accept new technologies such as artificial intelligence methods.

The following section discusses generic techniques for modularity. This is followed by a multilevel, multimodal architectural approach that satisfies the functional requirements above. The final sections illustrate the ways in which the system architecture accommodates a spectrum of design and production capabilities.

10.3. MODULARITY AND SYSTEM ARCHITECTURE

Hierarchical and layered organizations [5, 7] are two generic techniques for modularization.

10.3.1. Hierarchies

The notion of hierarchies permeates the analysis and synthesis phases of engineered systems, since it arises in both domains. Hierarchical structures are found, for example, in the organization of industrial plants when analyzing manufacturing systems, as well as in functional trees generated when synthesizing designs.

The high-level hierarchical configuration for design and manufacturing is shown in Figure 10-1. Each of the lower-level blocks in Figure 10-1 may be further decomposed. For example, the "supporting equipment" block may include provision for a sentry robot to patrol the factory. This robot would in turn have its hierarchy of subsystems, as shown in Figure 10-2. The hierarchical structure of the communications network is shown in Figure 10-3.

10.3.2. Layers

Layering involves the stratification of system components into different slices or levels. A component in one layer can interact with only the adjacent layers, with which it communicates through a standard interface, as shown in Figure 10-4. The purpose of such an interface is to serve as a gateway that mediates all interlayer communication and to hide the structure of each layer from its neighbors.

A prime example is found in the levels of computer languages. Consider the progression of languages defined by a machine language M, assembly language A, procedural language P (e.g. FORTRAN or Pascal), and a nonprocedural language N (e.g. a program specification language). The sequence $<M, A, P, N>$ represents a total order among the languages, each of which represents a different layer. For example, the statements in a Fortran program need only interact with each other; the linkages among the corresponding assembly language statements are hidden by the Fortran-to-assembly interface (i.e., the compiler).

Another example of layering is found in the Open Systems Interconnection Reference Model adopted by the International Standards Organization as a telecommunicatons protocol. This system consists of seven levels ranging from the physical level, which pertains to signal interpretation, through the applications level, which deals with specific end-user packages.

An obvious analogy in the manufacturing environment is a system of description languages to specify the activities for production equipment, from machine tool control to workcell configuration. This sequence of languages might take the form of an assembly language for low-level numerical control, up to a high-level language for manufacturing cells. The layering strategy is discussed in detail in Section 10.4 in the context of the Design Advisor.

10.3.3. A Unified Architecture for Manufacturing

An integrated architecture for computer-integrated manufacturing is found in the explicit two-dimensional configuration consisting of both hierarchies and layers. The static facet of the architecture - such as physical equipment or software modules associated with different functional roles - is usefully envisioned in terms of a hierarchical organization. This view is complemented by the dynamic facet of software interactions, which is usefully implemented in a layered configuration.

A unified architecture for manufacturing admits structures of both the hierarchical and layered types. Both arrangements may be accommodated on the communication structure shown in Figure 10-3. The next two sections discuss in greater detail a number of key components that may comprise elements of such an integrated system.

10.4. DESIGN ADVISOR

In the conventional design scenario, an engineer relies on personal knowledge and experience to develop a number of prototype devices. In this task, he may be supported by computer simulation programs to aid in understanding and selecting the appropriate design parameters. The existing approach, however, suffers from the following limitations:

1. There is no guarantee that this approach will yield the optimal solution or even a rational solution.

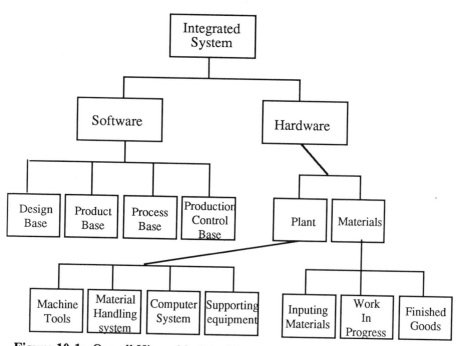

Figure 10-1: Overall Hierarchical Architecture for Design and Manufacturing

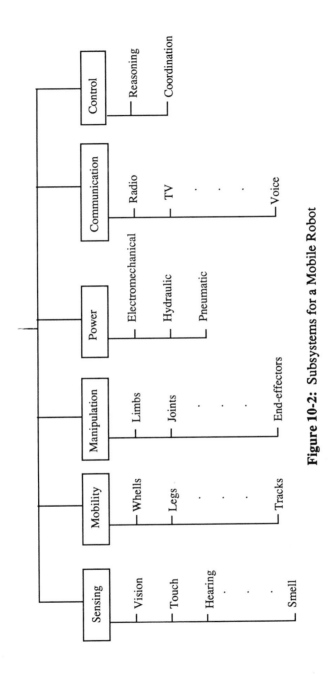

Figure 10-2: Subsystems for a Mobile Robot

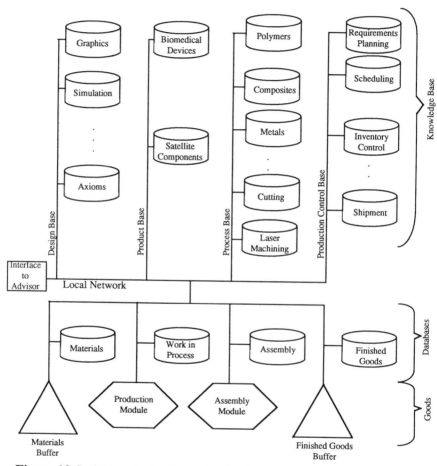

Figure 10-3: Network Configuration for Computer Integrated Design and Manufacturing

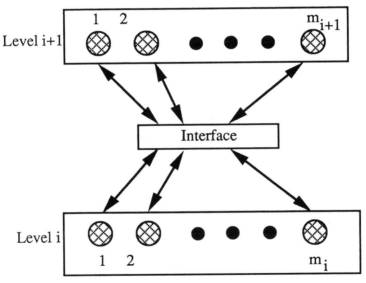

Figure 10-4: Adjacent Levels L_i and L_{i+1} for a
Layered Configuration [5]

2. The intuitive knowledge available to the experienced engineer cannot be generalized for application to different situations.

3. Such personal knowledge and experience cannot be readily transmitted to others.

An alternative approach is to develop rational decision procedures for satisfying the functional requirements, rather than the optimization of existing design parameters. Less research has been done for the synthesis phase than for analysis, perhaps because the synthesis of new designs is inherently more difficult, often defying conventional analytical and empirical techniques.

For these reasons, a new design methodology is required which meets the following requirements:

1. It allows for the exploration of new ideas with minimal risk.

2. It shortens the product development lifecycle.

3. It ensures the superiority of the end product.

An *axiomatic design* approach is being developed with these ends in mind [1, 2, 11, 12]. This approach is based on the premise that there exist general decision rules for determining superior candidates in the design of hardware, software, and processes. The utility of the axiomatic approach has been demonstrated in a number of application areas ranging from robotic design to polymer processing.

A general set of guidelines has been proposed for this purpose in the form of *design axioms*. The axioms are a set of decision rules for evaluating alternative designs; these primary guidelines are supplemented by more specific decision rules called corollaries. For particular application areas, these combined rules may be augmented by guidelines that are even more specific. For example, a consultive system for designing microprocessor-based products may draw on a knowledge base incorporating specific heuristics. These ideas will be further amplified in the sections which follow.

A useful research prototype is currently being developed. This system will provide decision support for innovation (development of superior designs) and evaluation (testing of manufactured products). To ensure that the products of the research are applicable to realistic problems, the system is being developed in the context of specific products.

10.4.1. Design Axioms

The design process begins with a perceived need or problem which is then translated into a set of functional requirements. The *functional requirements* constitute the minimal set of independent specifications that completely specify the problem at hand. They are independent in two senses: (1) the specifications are not redundant, and (2) each specification must be satisfied in order to solve the problem.

Often, design problems are characterized by *constraints* that further limit the bounds of acceptable solutions. In some sense, constraints are of secondary importance because they leave much room for maneuvering. For example, a constraint on cost might specify that manufacturing cost must not exceed an upper limit of $50, while a constraint on reliability might require that the mean time to failure exceed 2000 hours.

The key to the design process is the mapping of a set of functional requirements into a corresponding set of *design parameters*. The design parameters are the critical attributes of a design that relate to the functional requirements.

The design axioms may be stated in the following form [11, 12]:

- **Axiom 1: Independence.** Maintain the independence of functional requirements. (An acceptable design is one that is feasible and uncoupled.)

- **Axiom 2: Information.** Minimize information content. (Of two acceptable designs, the one with less information is superior.)

The statements in parentheses are more precise versions of the axioms, in that they bring out the inherent assumptions [5]. Axiom 1 may be written in the logic programming language Prolog as:

```
acc(X)  :-
          feas(X),
          unc(X).
```

The interpretation is that a design X is acceptable if it is feasible and uncoupled.

The encoding of the second axiom is slightly more involved. For this purpose we must first define the construct *iff(X,Ifm)* which takes a design X and returns its overall information measure *Ifm*. Then Axiom 2 may be stated as

```
super(X,Y)  :-
          acc(X),
          acc(Y),
          iff(X,Ifmx),
```

```
iff(Y,Ifmy),
Ifmx < Ifmy.
```

This says that design X is superior to design Y if both are acceptable (by Axiom 1) and design X has less information content than Y.

10.4.2. Data Structures for Axioms 1 and 2

A general data structure to encode design data is $D(X,L)$ where D represents the name of a given attribute of design X, and L the corresponding list of specifications. When L is an empty set, we may write $D(X)$ rather than $D(X,[\ ])$. Consider, for example, the geometric information relating to a rectangular block. Let U, V, and W represent lengths along three spatial axes, with respective tolerances DU, DV, and DW. Then these specifications might be encoded in the form:

```
geomd(X, [U,DU,V,DV,W,DW]).
```

This is an example of a structure at the Basic or Data Level. We need a way to connect the Basic Level to the design axioms at the Primary Level. This intermediary role is the purpose of the Secondary and Tertiary Levels. For example, let *geomf(X,Geomm)* be the function that maps a block X into its geometric information measure *Geomm*. Then an entry in the Secondary Level would be:

```
geomf(X,Geomm) :- geomd(X, [U,DU,V,DV,W,DW]),
                  Geomm is (log(U/DU) + log(V/DV) + log(W/DW)).
```

where *log* denotes the base two logarithm.

10.4.3. Some Characteristics of Expert Systems

The advice of a human consultant is characterized, among other things, by the following attributes:

1. *Parsimonious interaction.* Based on a limited set of facts, the human expert can dramatically reduce the space of potential solutions consistent with those facts. In effect, the input facts allow for wholesale pruning of the search space of solutions.

2. *Incremental refinement.* Even a sparse set of input facts permits

the expert to sketch a potential solution or set of such. Additional facts allow for an evolutionary refinement of the preliminary solution until a sufficient level of detail and/or performance in the solution is obtained.

3. *Informative explanations.* An expert can provide explanations at varying levels of detail, as well as reason about causal mechanisms in the application domain.

These human characteristics should be emulated by software packages. To this end, a design advisor should incorporate these architectural features:

1. *Model-based reasoning.* A model or internal representation may be used to encode general characteristics of the application domain, as well as potential solutions. In this way, facts relating to aspects of the model can lead to a chain of inferences concerning related aspects. Further, a model of the domain facilitates the generation of informative explanations both in terms the detail of explanation as well as causal inferences.

2. *Blackboard architecture.* A blackboard provides a common area in which to develop solutions incrementally, much as a ship under construction in a shipyard. The blackboard serves as a general notice board for different components of an advisor, which may contribute to the evolutionary development of the solution.

These features are incorporated in the design advisors described in the sections to follow.

10.4.4. Axiomatics Design Advisor

The system architecture for the Axiomatics Advisor consists of two Modules, as shown in Figure 10-5. The Kernel incorporates domain- independent knowledge while the Application Module handles domain-specific knowledge.

The Primary Level of the Kernel consists of the Design Axioms encoded in software. The Basic Level of the Application Module consists of two components: the Application Base, containing rules specific to the domain; and the Case Base, a workspace for case-specific information. For example, an advisory system for designing intelligent devices may contain information such as the clock rates for different microprocessors. In contrast, the Case Base contains information such as the functional requirements for a specific product under consideration.

The Advisor incorporates a User Interface that provides the user with friendly interactions and explanation facilities. Any problems which arise during a consultation are handled by the Error Handler.

The Secondary and Tertiary Levels represent intermediate stages to interconnect the Primary and Basic Levels. As depicted in Figure 10-6, the Secondary Level contains invariant knowledge. For example, it contains procedures that specify how to calculate geometric information, given appropriate data specifications. The Tertiary Level (Figure 10-7), in contrast, specifies whether or not geometric information is relevant for specific aspects of the designs to be considered. The Application Base contains knowledge pertaining to the product domain; this is illustrated in Figure 10-8.

The Primary and Secondary Levels together constitute a model of the axiomatic design approach. The Tertiary Level incorporates a model of the application domain. The Case Base includes provision for a blackboard in which to evolve the solution incrementally (Figure 10-9). The blackboard consists of (i) a Fact Board to keep track of input assertions such as the specific set of functional requirements and known observations, (ii) a Solution Board to display the evolving design, including assertions deduced from the Fact Board, and (iii) a Meta Board to post information such as the level of explanation appropriate to a particular user.

The functional interpretation of the Design Advisor is shown in Figure 10-10. The knowledge pertaining to a specific design problem is contained in the Case Base. The Application Base serves as a change agent which evolves the design according to the knowledge on the Fact Board and under guidance from the Tertiary Level.

As mentioned previously, the Secondary and Tertiary Levels act as the translator which mediates the Primary Level and the lower levels. The Primary Level serves as the ultimate judge; at each iteration of the evolving design, it determines whether a superior design has been attained or further work is required.

This type of expert system serves two roles: it acts both as a guide to direct the designer's thinking, and as a tutor to remind him of axiomatic reasoning. An example of the second function occurs when the system reminds a forgetful user that the design parameters should be as numerous as the functional requirements.

In general, a computer system cannot deduce the functional requirements of a product or process merely from a list of specifications in a database [3]. However, this general limitation does not preclude the development of a system to evaluate designs in restricted domains of application, once the relationships among functional requirements and potential design parameters are clarified. The Axiomatics Advisor, as an expert system shell, may be tailored to different applications by simply changing the contents of the Application Module.

The next section illustrates how the Advisor may be tailored and used in a specific application relating to the design of intelligent products.

10.4.5. Smart Products Advisor

Computer technology allows for the transfer of simple information-processing activities from people to machines. One result is the advent of intelligent products incorporating microprocessors. These micros provide intelligence by incorporating conventional algorithms as well as knowledge bases. Over the next decade, we may expect such knowledge bases to have the equivalent power of current systems which exhibit perhaps a thousand rules. This phenomenon foreshadows the need for a consultive system to help designers build intelligent products.

The overall structure of the Smart Products Advisor (SPA) is as shown in Figure 10-5 for the Axiomatics Advisor. We may be more specific, however, about the substructure of the Application Base. This knowledge base consists of the following four subsystems:

1. Hardware Base, containing information such as performance and price data for microprocessors.

2. FR Base, incorporating a list of potential functional requirements.

3. DP Base, containing a list of potential design parameters.

4. Guide Base, incorporating auxiliary decision rules specific to the design of intelligent devices.

The FR and DP Bases contain domain-specific information on candidate sets of functional requirements and design parameters. These databases may be used by the Guide Base to offer helpful suggestions or to validate the consistency of user input. For example, the Guide Base may be used to inform the user that a certain microprocessor has insufficient memory to fulfill a particular software requirement. Another illustrative use of the Guide Base is to offer suggestions on how certain design parameters, either singly or in combination, might be used to satisfy specific functional requirements.

Suppose that the functional requirements for a microprocessor system are given in terms of

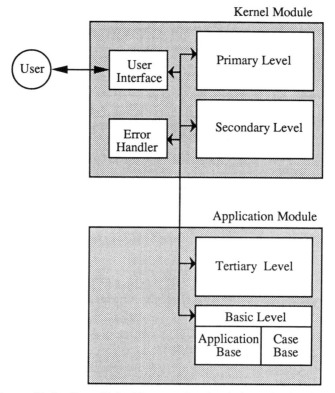

Figure 10-5: Overall Architecture for the Axiomatic Advisor [3]

Secondary Level

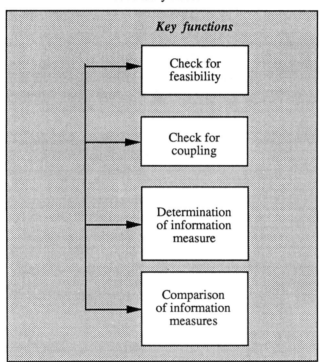

Figure 10-6: Components of the Secondary Level

Tertiary Level

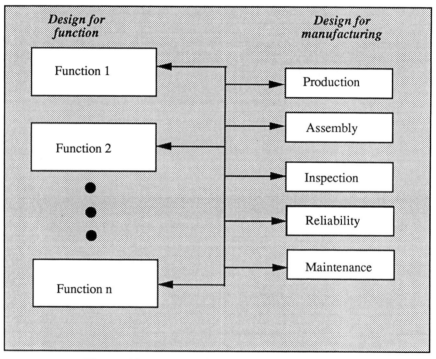

Figure 10-7: Components of the Tertiary Level

Application Base

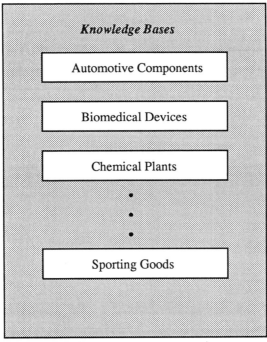

Figure 10-8: Components of the Application Base

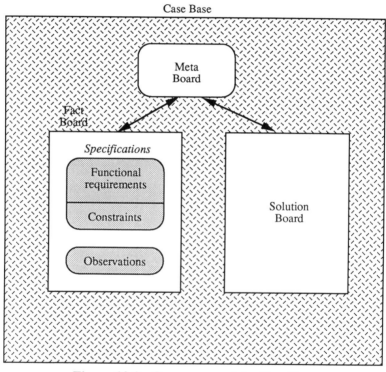

Figure 10-9: Components of the Case Base

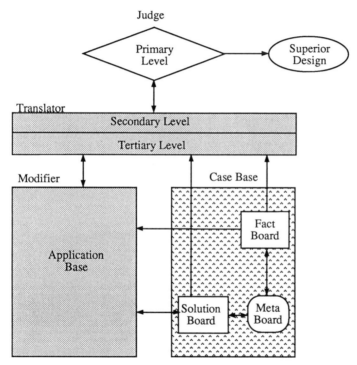

Figure 10-10: Functional Interpretation of the Design Advisor

1. System response time, R.

2. Knowledge base size, K.

The response time pertains to the rapidity with which the system reacts to new input, while the knowledge base size refers to the number of rules encoded in the software.

In addition, assume that the designer suggests the following set of potential design parameters:

1. Cycle time, C.

2. Word size, W.

3. Number of connector pins, N.

4. Random Access Memory (RAM) size, M.

The Advisor would first explain that the design parameters are redundant, since they exceed the number of functional requirements. The system might then suggest a composite measure of processing power, defined as $P = N * W / C$. This quantity affects response time R but not the knowledge base size K.

Obviously, the memory size M affects the knowledge base size K. But does M affect the response time R? That depends on the specific software strategy used. In particular, the use of a hashing strategy to seek entries in a look-up table will imply that an increase in M would decrease R. But the use of a hashing algorithm will result in an uncoupled design. In accordance with the discussion in Section 10.4.1, the system is uncoupled and therefore admissible by Axiom 1.

The application of Axiom 2, on the other hand, would suggest a design with minimal information requirements. For example, it would indicate that memory size M should be no larger than that required to fulfill the functional requirements.

10.5. DESIGN & MANUFACTURING ADVISOR

10.5.1. Integration of Design and Manufacturing

The amount of information available in an organization expands with the size and complexity of the organization. Hence, different departments in large organizations tend to be compartmentalized. For example, the marketing people may request products from the design department based solely on marketing studies, without due consideration of the difficulty of meeting certain specifications, or taking advantage of newly developed technologies.

In a similar way, the design engineers may create products without adequate appreciation for the difficulty of producing certain components or characteristics. For example, the relaxation of a tolerance specification by 5% may permit the use of an alternative production process which decreases manufacturing cost by 30%. If such leveraging information were to be readily communicated, the overall productivity of the organization would rise dramatically. Despite concerted efforts in the past to enhance interdepartmental cooperation, however, the pressures of short-term deadlines have discouraged such cooperative efforts. The use of computers, however, holds promise as a medium for transmitting such leveraging information across organizational boundaries.

10.5.2. Design & Manufacturing Advisor

A decision support system can aid a designer in producing designs that are easily manufactured. By storing the knowledge of production engineers in a knowledge base, it is no longer necessary for the manufacturing specialist to become directly involved in the design phase. The designer can be guided in his creative efforts by the computer system, which draws on its knowledge base to make appropriate suggestions. A prototype for such a decision support system is currently under development [1, 3].

The Design & Manufacturing Advisor (DMA) is a generalization of the Axiomatics Advisor. The system is intended to assist a human designer in the following tasks:

1. Transforming the perceived need into a quantitative set of functional requirements.

2. Exploring alternative feasible designs.

3. Studying the manufacturing implications of the design by (a) examining alternative manufacturing processes, and (b) determining their impact on the performance and cost of the final design.

4. Selecting the best design and production procedure for the case.

5. Issuing the production order to fabricate the design.

Perhaps the simplest of these functions is the fifth item, which is a relatively well-defined task once the requisite manufacturing processes have been identified.

The system architecture to support the DMA is as shown in Figure 10-3. The DMA is designed to interact with databases through a local area network. The *software* subsystem consists of four main knowledge bases: (a) Design Base, including packages such as graphics and simulation, (b) Product Base containing knowledge of specific products, (c) Process Base incorporating knowledge such as composites or metals processing, and (d) Production Control Base for such tasks as scheduling.

The *hardware* components consist of the following four modules: (a) Raw Materials Buffer, (b) Production Module, (c) Assembly Module, and (d) Finished Goods Buffer. No module is shown for inspection, since each stage can perform its own quality control functions. The system configuration shown here is obviously a canonical high-level view of the factory. When greater detail is required, each component may be expanded further in the spirit of the IDEF modeling technique developed under the Air Force ICAM Program. Each hardware module can maintain its own database. For example, the Production Module is served by a Production Database that maintains static information such as the type and number of NC machines, as well as dynamic information such as the job numbers of the workpieces in the Module. The user interacts through a friendly interface consisting of icons, dynamic windows, and natural language.

The purpose of this knowledge-based system is to dramatically reduce the development time from product conception to final fabrication. The architecture as described is useful as a conceptual scheme, and its utility need not await the translation of the myriad items of informal knowledge into software. In fact, given the rapid accumulation of knowledge in manufacturing, the knowledge base will continue to evolve. The conceptual structure is useful even for partial implementations, as illustrated in the prototype system for designing turbine disks for aircraft engines [4].

10.5.3. Integration of Design, Manufacturing and Maintenance

The goal of the Design & Manufacturing Advisor is to integrate the design and fabrication functions. An extension for the future is to integrate forward into the post-production phase. In particular, an intelligent design advisor should be able to inform the user of the availability, reliability, and serviceability of each design in addition to its cost and production implications. The system would incorporate the modeling of manufacturing processes with sufficient clarity for accurate simulation. Such information may be combined with a knowledge of product attributes such as geometric tolerances, to develop useful models of service life characteristics.

10.6. ACKNOWLEDGEMENT

I would like to thank Mark Silvestri for thoughtful suggestions which have helped to improve the presentation in this paper.

10.7. BIBLIOGRAPHY

[1]   Kim, S. H. and Suh, N. P., "Application of Symbolic Logic to the Design Axioms," *Robotics and Computer-Integrated Manufacturing,* Vol. 2, No. 1, pp. 55-64, 1985.

[2]   Kim, S. H. and Suh, N., "On an Expert System for Design and Manufacturing," *COMPINT '85, ACM and IEEE/Computer Society, Montreal,* pp. 89-95, September 1985.

[3]   Kim, S. H. and Suh, N. P., "Mathematical Foundationas for Manufacturing," *Journal of Engineering for Industry,* Vol. 109, No. 3, pp. 213-218, 1987.

[4]   Kim, S. H., Hom S. and Parthasarathy, S., "Design and Manufacturing Advisor for Turbine Disks," *Robotics and Computer-Integrated Manufacturing,* Vol. 4, No. 3/4, pp. 585-592, 1987.

[5]   Kim, S. H., *Mathematical Foundations of Manufacturing Science: Theory and Implications,* unpublished Ph.D. Dissertation, School of Management and Dept. of Mechanical Eng., M.I.T., Cambridge, MA 02139, 1985.

[6]   Kim, S. H., "Frameworks for a Science of Manufacturing," *North American Manufacturing Research Conference,* Minneapolis, pp. 552-557, May 1986.

[7]   Kim, S. H., "Sources of Industrial Productivity and Strategies for Research," *Proc. International Conference on Production Research, Cincinnati,* Ohio, pp. 117-123, August 1987.

[8]   Kim, S. H., *Designing Intelligence,* Oxford University Press, New York, 1990.

[9]   McLean, C. M., "A Computer Architecture for Small-batch Manufacturing," *IEEE Spectrum,* Vol. 20, No. 5, pp. 59-64, May 1983.

[10]   Spur, G., "Growth, Crisis and Future of the Factory," *Robotics and Computer-Integrated Manufacturing,* Vol. 1, No. 1, pp. 21-38, 1984.

[11]   Suh, N. P., Bell, A. C., and Gossard, D., "On an Axiomatic Approach to Manufacturing and Manufacturing Systems," *Journal of Engineering for Industry, Trans. of ASME,* Vol. 100, No. 2, pp. 127-130, May 1978.

[12]   Suh, N. P. et al., "Optimization of Manufacturing Systems Through Axiomatics," *Annals of the CIRP,* Vol. 27, No. 1, pp. 383-388, 1978.

[13]   Wernecke, H. J., "Conflicting Objectives in Designing Market-Oriented Production Structures," *Robotics and Computer-Integrated Manufacturing,* Vol. 1, No. 1, pp. 51-60, 1984.

Chapter 11
DUAL DESIGN PARTNERS IN AN INCREMENTAL REDESIGN ENVIRONMENT

Mark J. Silvestri

ABSTRACT

Many real world product creations such as automobiles, aircraft and appliances are the result of incremental modifications to previous designs. Architecture decisions in computer support systems for these product creations inherently influence the actual computer system behavior. The dual design partner paradigm described in this paper purposefully supports two competing system behaviors. One expert machine, the *stabilizer*, resists change and always presents a conservative hypothetical model of the product. The other expert machine, the *innovator*, strives for well-calculated and justified alternative hypothetical models of the product. The importance of this paradigm is that it provides a model to support a check and balance mechanism for product creation. In addition to the basic description of the dual design partner paradigm, this paper describes an approach to implementation that utilizes the blackboard architecture as a keystone technology. The most pragmatic element of the implementation philosophy is the blending of database management system and AI technology.

11.1. INTRODUCTION

Many new products are actually enhancements to existing versions of a similar product. Even relatively unique products are typically refinements of prototype products. Engineering, design and manufacturing environments that can be characterized by a predominance of incremental redesign, yet require occasional well calculated innovations, present a unique opportunity for experimentation in knowledge-based expert systems for product fabrication. Because the deliverable product will generally deviate only slightly, specifications, drawings, logistic support plans and pictorial process plans including all machinery, assembly and packaging information will also deviate only slightly. This inherent resistance to change of the design environment facilitates the orderly acquisition of credible and reliable knowledge about the product. That is, in a practical sense, it is easier to describe and document a static model than a dynamic model. Nevertheless, there are always changes, and occasionally, catastrophic changes when significant innovation is introduced into the product configuration.

This clash between the need to standardize the product process in order to maintain high productivity and the need to innovate in order to maintain a technological edge is commonplace in the traditional world of engineering, design and manufacturing. It can be viewed as the check and balance mechanism for product creation. An effective model for a computer design partner would take into account both goals.

11.2. THE DUAL DESIGN PARTNER PARADIGM

The dual design partner paradigm calls for two competing expert machines which would co-exist in this predominantly incremental redesign environment. One expert machine, the stabilizer, would resist change and always present a conservative hypothetical model of the product. Its role would be the facilitator, getting the job done in a fashion as close as possible to the way all previous products have been done. The other expert machine, the innovator, would strive for well calculated and justified alternative hypothetical models of the product. The brief list in Table 11-1 provides a profile of each expert machine.

In general, the innovator expert machine requires less explicit rules and control information than the stabilizer expert machine in order to operate. However, the results of the innovator are potentially more error-prone than the stabilizer due to uncertain knowledge. As a consequence, this leads to the requirement for frequent feedback to the innovator from the engineering, design and manufacturing community. Both expert machines employ generalization which inherently improves in accuracy as the sample size increases.

A unifying knowledge construct that must be captured and made available for analysis to both expert machines is the part-whole abstraction, or assembly tree. This assembly tree is polymorphic throughout the stages of the product process. That is, the various functional perspectives in the product process of the product

Table 11-1: Profiles of Expert Machines

STABILIZER	INNOVATOR
Uses more algorithms than heuristics	Uses more heuristics than algorithms
Algorithms tend to be those used to run well-defined subprocesses	Algorithms tend to be those used to support the generation and validation of new hypothetical product models

Familiar strategies are:
    Copy and Edit;
    Replay;
    Inference using explicit rules created by humans or induced rules approved by humans
	Familiar strategies are:
	    Automated knowledge acquisition and computer learning;
	    Conceptual clustering /Knowledge organization;
	    Reasoning by analogy;
	    Introspection

Typically completes entire tasks and usually without error	Typically completes only a portion of tasks; requires frequent feedback to refocus; results are more error-prone
Operates with approved knowledge	Operates with uncertain knowledge
Compensates for human flaws like fatigue in repetitive tasks	Compensates for human flaws like early termination of the evaluation of alternative hypothesis
Dominates the product process	Only occasionally has impact on the product process
When faced with competitive goals, this expert machine tends to order goals as was previously ordered in other product histories	When faced with competitive goals, this expert machine tends to order goals to defer commitments to impose the fewest restrictions on the form of the design solution

components require a transformation of the part-whole abstraction even though the constituent information is the same. For example, a conceptual designer may create a part-whole hierarchy based on subassembly behavior, while a manufacturing engineer may reorganize the components based on assembly ease. When a product is in the field, its approved instantiated assembly tree represents an achieved goal for the product engineering, design and manufacturing organization. Associated with the tree and each node and leaf of the trees are the specifications, drawings, logistic and process plans of the product and its assemblies, subassemblies and components. Collections of the instantiated assembly trees which are approved, plus their associated information become the experiential knowledge base for the product engineering, design and manufacturing organization.

Both the innovator and the stabilizer require feedback from a number of people involved in the product process. Figure 11-1 pictorially represents the engineering, design and manufacturing community in frequent contact with the stabilizer and innovator expert machines, which in turn negotiate with each other.

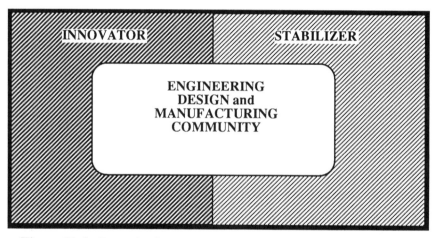

Figure 11-1: Pervasive Innovator and Stabilizer Design Partners in the Engineering, Design and Manufacturing Community

As mentioned previously, each approved instantiated assembly tree represents an achieved goal for the engineering, design and manufacturing organization. In practice, not every component in an approved assembly is a marvelous achievement. One fundamental item of information that will be stored with each approved component is its *fitness profile*. At a minimum, this would be a qualitative judgement made by the marketing, engineering, design, manufacturing and field service approvers regarding the fitness of a given component from their perspective. For example, an approver would supply a statement like "poor" or "good" or a number from 1 to 10. In a more robust scheme, each approver class would provide a quantitative profile of how well the component satisfies its specifications. Over time this checking could be automated allowing the approvers to concentrate on creating and refining specifications.

This fitness profile identifies target components on which the innovator can focus its activity. It also becomes the basis for the stabilizer's recommendation to the engineering, design and manufacturing community to accept or reject a new approach being recommended to that same community by the innovator.

11.3. RELEVANT WORK

Many of the ideas presented in the dual design partner paradigm have been the subjects of theoretical AI research for the past 15 years. The blend of two dichotomously opposed expert machines, a stabilizer machine and an innovator machine, superimposed over an environment characterized by a predominance of incremental enhancements to existing products, is unique. As with information modeling, the full definition of a stabilizer process agent and an innovator process agent will provide insight into better ways of managing the interaction and contribution of human stabilizer and innovator experts in the engineering, design and manufacturing domains.

A decision support system model which is similar to the dual design partner concept presented here is the "left-brained versus right-brained" paradigm [4]. Perhaps the major difference is that not only does the stabilizer expert machine rely heavily on left-brained processing, numerical and analytic processing, but the goal solving strategy of the stabilizer's inference engine tends to resist perturbation of the system. Its measure of effectiveness lies in the generation of correct solutions. The innovator's measure of effectiveness is that the joint human-machine performance is greater than that of the human or machine element alone. Woods [19] states that for effective cognitive coupling of humans and machine, the computational technology should aid the user in making a decision, not in recommending solutions. With partial solutions, frequently the user will have to rethink the problem anyway to determine why the computer failed; with complete solutions, the user base may, over time, have difficulty discriminating correct from incorrect machine solutions due to loss of experience. This project will probably demonstrate that an innovator expert machine that provides a combination of both solutions and decision aids is the most valuable design partner. Some decision support system researchers conjecture that a total decision support system would supply not only interfaces to left-brained and right-brained modules, but also an interface to modules which aid in the transition from qualitative decision problems to quantitative decision problems. The innovator differs from a purely right-brained system in that it relies heavily on analysis to verify its fitness. That is, it will utilize deterministic tools to support innovation. The transition from qualitative to quantitative judgement all occurs within the innovator expert machine.

On the general subject of ideas and issues in AI research relative to design, Mostow [11] and Chapter 1 of this volume provide a comprehensive review. Relative to expert systems capturing the intent of designers' actions, Mitchell et al [10] and Tomiyama et al [17] both provide some ideas on the subject. Other interesting research includes the use of common sense knowledge to overcome knowledge acquisition bottlenecks [9], and conceptual clustering as a knowledge organization technique [2].

Several more important references will be cited in the sections that follow. Many more references could be generated which share ideas with some facet of the dual design partner paradigm (see section 11.2). The important distinction in this work is that the specific problem domain and the blend of expert machines provide a practical environment for realizing tangible gains in the whole product process.

11.4. REFINING THE ABSTRACTION

Thus far, the dual design partner system has been described as an abstract notion. The idea of two competing expert systems capable of facilitating the entire engineering, design and manufacturing process, although intriguing, seems unrealizable.The remaining sections will focus on increasing the realness of the dual design partner system.

To summarize thus far: the motivation for the dual design partner system is a check and balance mechanism for product process automation. The innovator and stabilizer behavior is prevalent in the traditional engineering, design and manufacturing world and the goal of this work is to model this behavior in two distinct expert machines in an attempt to emulate and enhance the manual procedure.The primary role of the innovator and stabilizer machines is as advisors and facilitators for the product process; hence they should communicate to the advisees in their own perspective. The secondary role is to produce an actual product, or at least the drawings, specifications, documentation, numerical control instructions and other symbolic artifacts of the product process.

This article defines the dual design partner paradigm. Much work is still required to completely validate this approach. Nonetheless, some essential facilities -- a CAD/CAM/CAE modeling and drawing system, a large scale data management system and a group technology and process planning system -- are already commercially available through Computervision's product offerings. Computervision has an installed customer base in the thousands from which to draw practical experience in using CAD/CAM/CAE for real world product fabrication. Waters [18] in reviewing KBEmacs (which was a subproject within the Programmer's Apprentice project), stated that the "pragmatic non-AI problems tended to swamp all other considerations." Computervision has invested labor-centuries on very pragmatic issues.

The next section will show that the dual design partner system can be scoped and is achievable. A general architecture will be presented. This architecture will be decomposed two levels deep, and the major work areas will be identified.

The implementation section will identify a few key methodologies integral to answering how the dual design partner system will be built.

Finally, a discussion will be provided incorporating some of Computervision's products in order to reflect the foundation assumed for this project.

11.5. ARCHITECTURE

11.5.1. Dual Design Partners Process Decomposition

A convenient method for describing a complex system is to identify its component processes while also identifying data flowing into, out of and between each process. The system and subsystem boundaries are thus well defined and a holistic system view can be documented. Figure 11-2 is a process decomposition diagram for the dual design partners system. For this diagramming technique, a box represents a process. The lines represent data flowing into and out of a process. Each line is labeled with a data description. The arrow heads represent direction of flow.

Information and knowledge flow into the dual design partner system from two sources: feedback from the engineering, design and manufacturing community; and logs and recordings from various computer monitors external to the dual design partners system.

The raw information and knowledge flowing into the dual design partners system is ultimately transformed into advice and product with some degree of innovation and a complementary degree of stabilization.

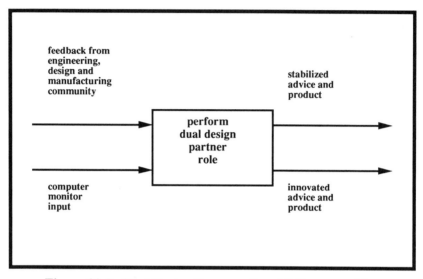

Figure 11-2: Dual Design Partners - Level 0 Decomposition

The dual design partners system is assumed to have an innate and self-activating ability to perform its function. A minimal product database, a knowledge base, a minimal process database, and a meta-knowledge base, are components of such a system.

In order to perform the dual design partner role, subprocesses within the dual design partner system must perform the following three functions:

- Acquire Information and Knowledge (1.0)
- Analyze Information and Knowledge (2.0)
- Advise Engineers, Designers and Manufacturers (3.0)

Figure 11-3 provides a level 1 decomposition of the system. An internal feedback bus provides an introspection capability. That is, self-evaluation is an additional input to the system.

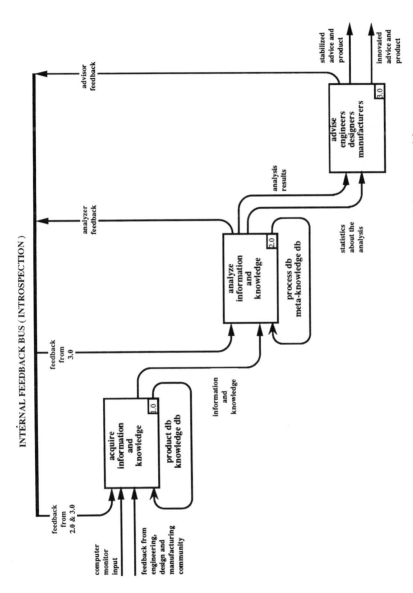

Figure 11-3: Dual Design Partners - Level 1 Decomposition

The distinction between product information and knowledge is more qualitative than quantitative. Indeed, few researchers feel that exploring the difference between "knowledge" and "data" aids in clarifying KBMS issues [1]. They were identified as separate items in order to indicate clearly that a large volume and variety of information is acquired. Since the distinctions are internal to the acquisition and the analysis processes respectively, only at much lower levels of decomposition where knowledge organization is occuring do the distinctions need to be formally defined. Process information is considered meta-knowledge because through a complete description of process, instances of product knowledge can be generated.

Acquire Information and Knowledge (1.0) process can be decomposed further as follows:

- Receive information and knowledge from computer monitors.
- Receive information and knowledge from engineers, designers and manufacturers.
- Receive analyzer and advisor feedback.
- Classify information and knowledge.
- Manipulate product and knowledge database.
- Send information and knowledge for analysis.

The Analyze Information and Knowledge (2.0) process can be decomposed further as follows:

- Receive information and knowledge for analysis.
- Receive advisor feedback for analysis.
- Perform stabilizing analysis.
- Perform innovating analysis.
- Manipulate process and meta-knowledge database.
- Send analysis results.
- Send analysis statistics.
- Send analyzer feedback to feedback bus.

The Advise Engineers, Designers and Manufacturers (3.0) process can be decomposed further as follows:

- Receive analysis results.
- Receive analysis statistics.
- Create and structure innovated advice and product.
- Create and structure stabilized advice and product.
- Send advisor feedback to feedback bus.
- Present innovated advice and product.
- Present stabilized advice and product.

11.5.1.1. Level 2 architectural description

Acquire Information and Knowledge.
A blackboard implementation [12, 13] will be used for collecting information and knowledge from: computer monitors; engineers, designers and manufacturers; and the dual design partner system's own analyzer and advisor feedback. After receiving input,the first action of the dual design partner system will be to classify the information and knowledge.

Classification involves determining the input type, quality and relevance to the solution state.In general, input from the computer monitors will have a low signal-to-noise ratio; feedback from the engineers, designers and manufacturers will be assumed 100% reliable and germane; and feedback from dual design partner introspection will carry with it a confidence factor. An example of computer monitor input is the observation that a CAD designer filed a CAD model at the end of a work session. Noise, in this case, would be the log entries that a CAD designer instantiated a particular object into his model and then moments later deleted it. Tolerance information, assembly envelopes and available machinery as of some effective date are examples of feedback from the engineers, designers and manufacturers. Deviation from a standard process plan is an example of introspective feedback. If the input is classified as relevant to the solution state, then receipt of the information and knowledge is posted as an event to evoke a reaction from the innovator and stabilizer analyzer processes.

A large, multi-gigabyte, product/knowledge database will always be available for analyzer use. Included in this database are items like assembly part-whole hierarchies, specifications, drawings, logistic support plans and pictorial process plans including all machinery, assembly and packaging information. Inquiries to the product/knowledge database will be processed through this module. Relevant incoming information and knowledge will also be stored in the database via this module.

Sending the information and knowledge for analysis is preceded by the

"Analyze Information and Knowledge" module recognizing that a significant event had occurred.

Analyze Information and Knowledge. Once a significant event is posted, information, knowledge and advisor feedback is received for analysis onto the blackboard. This influx of data is not random but rather it is associated to one or more structured activities on the blackboard.

Two solution hierarchies are active at all times on the blackboard: the conservative product hypothesis and the alternative product hypothesis -- a collection of incomplete alternative hypotheses. At any point, the conservative product hypothesis most accurately reflects the product and the product process. Within the solution hierarchies are levels of product abstraction. For example, at the lowest level are graphic primitives which provide model and schematic representations. These primitives are clustered to form a feature at the next higher level. Not only is there physical model information within the feature abstraction, but also steps in a process plan to create that feature as well as the standard documentation associated with it. The next higher abstraction is at the component level, then the assembly, and finally the product with the complete set of drawings, specifications, process plans and other symbolic artifacts of the product process.

On the blackboard the solutions are incrementally advanced. Frequent feedback from the engineers, designers and manufacturers is the guiding force of the process. The system provides a combination of both solutions and decision aids as a valuable design partner.

The two solution hierarchies are supported by two separate analysis behaviors. The dual design partner system will not use a large collection of rules, each with equal weight, in order to create innovator and stabilizer behavior. Brown and Chandrasekaran (Volume I, Chapter 7) in their treatment of routine design successfully use a hierarchy of specialists to focus the problem solving behavior of their expert system. Each specialist is responsible for the design of some portion of a component. Specialists use one or more plans to accomplish their design activity. Each specialist can either perform a task directly or call subordinate specialists to perform a task. In this model, design refinement is viewed as the cooperation of specialists that are executing plans, making design commitments for mutually dependent attributes and handling failure as locally as possible. Their model of routine design imposes the restrictions that the structure, part-whole hierarchy, is fixed and that the methods of completing the design artifact are known. Waiving these restrictions, a constrained specialist network seems particularly relevant to the dual design partner system.

Most CAD/CAM/CAE users work hard to standardize their product process. They create scripts and parameterized macros to help fill the gaps between vendor product and specialized product process needs. Phillips [14] created an automatic procedure for drawing a tool layout for a Vertical CNC

Turning/Boring Machine which has two independently controlled turrets and automatic tool change capability. He achieved a 5:1 to 10:1 production ratio increase. This automation was achieved by using the CADDS4x CAD/CAM/CAE package, by standardizing the process and by developing a few macro programs. Stored in a process/meta-knowledge database, these scripts and macros can be utilized by the stabilizer expert machine to perform specialist activity in specific areas of the product process. Similarly, once a process plan is created for a component, it is stored as a standard plan. The standard plan represents the best way to perform that particular task in that particular shop.When creating a new plan for a similar component these standard plans are used in copy/edit mode to facilitate the process planning. Not only are the process plans created more quickly utilizing the experiential knowledge of a particular installation, but the machinists on the manufacturing floor work with less ambiguity in their plans since a given task is described using the same vocabulary and specifications.

Targeting components for change can be accomplished by: obtaining specific feedback from the engineers, designers and manufacturers via constraints; identifying low fitness components by an inquiry to the product database; identifying low fitness components caused by the present evolution of the conservative product hypothesis; and, identifying a higher fitness for a substitute component in either the current conservative product hypothesis or in one of the alternative product hypotheses. In the case of a low fitness component, new components can be generated by comparing design plan preconditions and executing that plan (e.g., as described by Navinchandra in Volume II, Chapter 3). Where the preconditions do not have an exact overlap, an inexact analogy can be created by plan abstraction. The more inexact the analogy, the more general the abstract plan becomes which leads to more details that need to be filled in to solve the problem. A large process database containing many plans each with stored preconditions provides for variability in unique solutions.

For design problems in which some well structured compositional methodology was used for previous solutions, shape grammars [16] seems to be another useful innovator tool. This technique was used for generating three new Frank Lloyd Wright prairie house designs [7]. In a separate study, this technique was used to generate paintings in the style of the contemporary artist Richard Diebenkorn [6]. Although the use of shape grammars does not appear to be useful in a general problem solving environment, a specialist could certainly be encoded to use this technique for a specific task.

During the analysis process, the innovator or stabilizer may require from the Acquire Information and Knowledge module either more information and knowledge or the qualification of received information and knowledge. The request results and the qualification feedback are communicated via the blackboard.

Ultimately, analysis results need to be sent to the "Advise Engineers, Designers and Manufacturers" module for presentation to the users. More accurately, the analysis results are posted on the blackboard. This triggers the Advisor knowledge source to create advice and product. The importance of a conceptually convenient user interface was underscored by Waters [18] relative to the KBEmacs effort where programmers still preferred to edit program text instead of the plans. He also noted that effective communication requires a shared vocabulary at a level higher than first principles. In short, if the dual design partner system cannot provide advice and product at a level that the product process people can understand and prefer to use, then the dual design partner system will be ignored. Since at least some of the advice will be in the form of a partial solution or contain some uncertainty, statistics from the analysis activity will accompany the analysis results.

Advise Engineers, Designers and Manufacturers. As implied by the feedback loop, dual design partner activity is an iterative process. Reaching convergence on the two solution hierarchies is a negotiation process. Although some of this negotiation can be driven by specialists in the innovator and stabilizer behavior networks, committing to constraints is still primarily the responsibility of the users.

One research activity which is on the same scale and even overlaps a good deal of the problem domain of the dual design partner system is the Intelligent Project Manager, Callisto [15]. Three project management tasks were researched: activity management, product configuration management and resource management. A primary accomplishment of this work is the use of constraint-directed negotiations in the specification and revision process. When inconsistencies or incompleteness is detected in the product knowledge base, constraints are isolated in the negotiation process and either relaxed or strengthened to resolve the inconsistencies or incompleteness. A user interface for Callisto which allows for sharing of defaults, decisions and assumptions and that helps express impacts in the complex project management process still requires further development.

Clearly, the task of presenting advice and explanation about changes and incremental advancement of a product hypothesis to a whole profile of user classes is an enormous task. A combination of standard report forms and graphics provides the communication vehicle. Model changes relative to existing products can be expressed using color overlays on graphics terminals. Using a macro language, the system will also graphically generate newly innovated product models.

In any case, user classes will be able to define standard methods to receive advice and product solutions and this will be an early priority in the dual design partner project. Ignoring this facet will doom the dual design partner system to eternal prototype status since this product process automator requires continuous feedback from many engineers, designers and manufacturers in the product process.

11.5.2. Architecture Summary

The dual design partners system was decomposed into three subprocesses: Acquire Information and Knowledge (1.0), Analyze Information and Knowledge (2.0) and Advise Engineers, Designers and Manufacturers (3.0). Components of each of these subprocesses were identified and a discussion was provided relative to several key working assumptions and methods in each.

A product and a process knowledge base are essential ingredients in the dual design partners model. Also, critical to this model is a feedback loop for iteration and refinement of the two solution hierarchies.

There are five areas in the level two decomposition that require a more detailed description. One limiting factor in providing this detail is the lack of empirical data. These topics will likely be the subject of a separate paper. The five areas are:

1. Classify information and knowledge

2. Perform stabilizing analysis

3. Perform innovating analysis

4. Create innovative advice and product

5. Create stabilized advice and product

Although some implementation strategy has already been elaborated, the next section will further detail this strategy.

11.6. IMPLEMENTATION

A fundamental component of the dual design partner system is the blackboard. Nii [12, 13] provides a thorough two part review of the blackboard model and working systems which are effectively using it. The types of applications that best utilize a blackboard implementation are those which have large solution spaces, employ opportunistic reasoning and deal in various levels of abstraction. The dual design partners system conforms to these three criteria. Dodhiawala et al [3] in their report on the first workshop on blackboard systems identify the issues relevant to this technology.

All blackboard systems have several things in common. The blackboard is a central repository for knowledge and the locus for advancement of the solution state. The solution is advanced incrementally by contributing knowledge sources which interact independently with the blackboard. The actual problem being solved can be divided into several levels of abstraction and the various levels can be operated on concurrently. An application specific problem solving behavior is employed to determine the focus-of-attention which can either be a knowledge source, or, a blackboard object or object hierarchy.

As mentioned in the architecture section, the dual design partners system employs two primary solution hierarchies-- the conservative product hypothesis and the alternative product hypothesis, both containing identical product abstraction levels. At the lowest level are primitive graphics and at the highest level is the whole product abstraction. The PDES Product Structure and Configuration Management model [5] with extensions for connection to a product process network will be the form of the whole product abstraction. These extensions are currently under study and will be the subject of a separate paper. The definition of a formalism to express constrained process and process specialist networks on the blackboard is of particular interest. The superimposition of these networks yields a plan for the product process.

The blackboard itself will be an object-oriented semantic network. The innovator and stabilizer are knowledge interpreters which can traverse the constrained networks. The blackboard is the integrated instance view of the product and process abstraction. It is the activity space. It is not the conceptual schema or the repository of all product and process knowledge in the engineering, design and manufacturing community. Various process specialists will be able to seek information from the relational databases and associated reference files which are the predominant storage models of existing product and process information. These process specialists must bridge the gap between a semantically charged blackboard representation and semantically deficient relational space. Taxonomies as an extension to the RDBMS environment [8] may prove useful for this purpose.

Certain blackboard manipulations will exploit the encapsulated behavior model of the object oriented paradigm thereby relieving the blackboard controller from some scheduling activities. For example, routine transformations of the product abstraction within a particular product phase can be accomplished by encapsulated behavior independent of controller scheduling. Refinement of the product abstraction will however be controller dependent.

There are two classes of knowledge sources external and four classes internal to the dual design partner system. Figure 11-4 depicts their relationship with the blackboard. The external knowledge sources are the many engineers, designers and manufacturers who will interact with the dual design partners system and the many computer monitors that log various user and system transactions. The internal knowledge sources are the innovator analyzer, the stabilizer analyzer, the information and knowledge classifier and the advisor. The innovator analyzer and the stabilizer analyzer interact with the alternative product

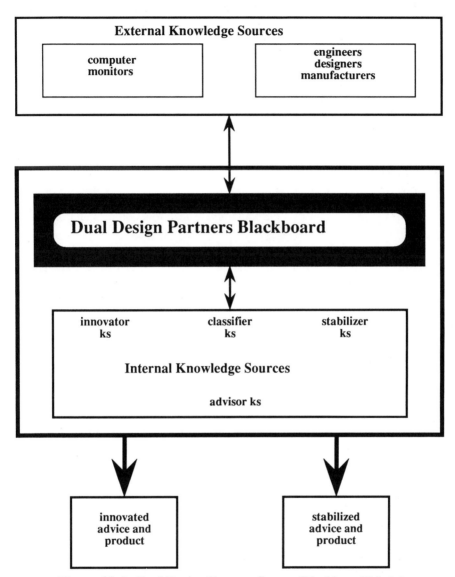

Figure 11-4: Dual Design Partners System Blackboard Model

hypothesis and the conservative product hypothesis solution hierarchies respectively. The information and knowledge classifier is a model-driven knowledge source which classifies relevant data for the blackboard. The advisor knowledge source continually attempts to facilitate the negotiation activity of the innovator and stabilizer in order to produce a convergence of the two primary solution hierarchies.

Not depicted in Figure 11-4, but already referenced in the architecture section, are global control data and a system control manager. That is, significant events are posted and an event scheduler sorts the priorities and determines the focus of attention. At least two primary sorting criteria are:

- Will the processing of a particular event advance a partial solution to a new level of abstraction?

- Will the processing of a particular event lead to a commitment on a mutually dependent attribute?

This discussion of the blackboard approach for the dual design partners system has partitioned the system into cooperating knowledge sources using a central information and knowledge repository with the main goal being the incremental advancement of a balanced product solution.

The next section will reference specific Computervision products and research activities to provide some insight into the data management foundation upon which the dual design partners system will be built.

11.7. COMPUTERVISION SYSTEM ENVIRONMENT AND TOOLS

Numerous advances in the Computervision hardware and software architecture as well as system configurability provide an environment for process-to-process communication for data management with relative ease regardless of the machine type or physical proximity of the process sharing machines. The system environment is rich in system resources:

- Networked SUN workstations with added accelerator boards running under a UNIX operating system.

- SUN server, DEC minicomputer and IBM mainframe environments.

The workstations support the interactive CAD/CAM/CAE environment and the server, minicomputer and mainframe computers support the on-line information storage and retrieval environment for the product and process knowledgebases. In addition, each expert machine resides on a separate SUN workstation to allow for parallel processing. All systems are networked together using TCP/IP. The applications insulate themselves from the network protocols and interprocess communication details via a distributed application network service, DAS.

The full power of the Computervision CAD/CAM/CAE system will be available to both expert machines. Two products are particularly appropriate for the data management aspects of the dual design partner system: PDM and CVpi II. PDM, Product Data Manager, provides a corporate vaulting mechanism. Entire projects can be controlled relative to access and security, revision levels, archiving and incremental backup and recovery. CVpi II, Computervision Project Integration, currently viewed as an advanced customer-tested prototype, provides a large scale multi-disciplinary model management capability. Individual CAD objects for ships, aircraft, offshore platforms and other macro-assemblies can be accessed and controlled for read or write privilege by geometric zone and discipline class. When change occurs in the model, all users of the objects in flux can be automatically updated with the changes in their own local workstation environment. Hence, CVpi II provides a centralized project model with support for a high degree of workstation local sufficiency.

Three languages will be *understood* and *utilized* by both the stabilizer and innovator expert machines:

- The Computervision CAD/CAM/CAE system interface

- A Computervision macro language, CVMAC

- The relational database query language, SQL

The term "understood" refers to the expert machines' ability to record the actions of the human engineering, design and manufacturing community, and then replay all or segments of the users actions, classify the objects created and begin to draw analogies to previous work as well as provide feedback to the users.

The term "utilized" refers to the expert machines' ability to generate CAD models, generate queries to satisfy an information void and provide themselves introspective services.

11.8. SUMMARY

A unique blend of two expert machines, one to stabilize and one to innovate, provides a check and balance mechanism for automated product production. A dual design partner system is proposed that will provide experimental data on the role of knowledge based expert systems in an engineering, design and manufacturing environment that is primarily characterized by incremental enhancements to existing products.

11.9. REFERENCES

[1]   Brodie, M., Balzer, R., Wiederhold, G. Brachman, R., Mylopoulos, J., "Knowledge Base Management Systems: Discussions from the Working Group," in *Expert Database Systems,* Kerschberg, L., Ed., The Benjamin/Cummings Publishing Co., Inc., pp. 19-26, 1986.

[2]   Cheng, Y., Fu, K. S., "Conceptual Clustering in Knowledge Organization," *Proceedings of the First Conference on Artificial Intelligence Applications,* pp. 274-279, 1984.

[3]   Dodhiawala, R., Jagannathan, V., Baum, L. and Skillman, T., "Workshop Report: The First Workshop on Blackboard Systems," *AI Magazine,* Vol. 10, No. 1, 1989.

[4]   Keen, P. G. W. and Scott Morton, M. S., *Decision Support Systems: An Organizational Perspective,* Addison-Wesley, 1978.

[5]   Kimura, K., "PDES Product Structure and Configuration Management, Ver.1.0," A PDES, Inc. Publication, 1989.

[6]   Kirsh, J. L., Kirsh, R. A. and Ressler, S, "Computers Viewing Artists at Work," in *Proceedings of Syntactic and Structural Pattern Recognition, NATO ASI SERIES,* Ferrate, G., Pavidis, T., Sanfeliu, A., Bunke, H., Eds., Springer Verlag, 1987.

[7]   Koning, H. and Eizenberg, J., "The Language of the Prairie: Frank Lloyd Wright's Prairie Houses," *Environment and Planning B,* Vol. 8, pp. 295-323, 1981.

[8]   Kostovetsky, A. and Silvestri, M., "A Taxonomy-Based Knowledge Representation Technique for Extending the Relational Model," *Proceedings of the Internationl Symposium of New Directions in Computing, IEEE 85CH2134-5*, pp. 72-79, 1985.

[9]   Lenat, D., Prakash, M. and Shepherd, M., "CYC: Using Common Sense Knowledge to Overcome Brittleness and Knowledge Acquisition Bottlenecks," *AI Magazine*, Vol. 6, No. 4, pp. 65-85, 1986.

[10]   Mitchell, T. M., Mahadevan, S. and Steinberg, L., "LEAP: A Learning Apprentice for VLSI Design," *Proceedings IJCAI85*, Los Angeles, CA., August 1985.

[11]   Mostow, J., "Towards Beter Models of the Design Process," *AI Magazine*, Vol. 6, No. 1, pp. 44-57, 1985.

[12]   Nii, P., "Blackboard Systems: The Blackboard Model of Problem Solving and the Evolution of Blackboard Architectures: Part I," *AI Magazine*, Vol. 7, No. 2, pp. 38-53, 1986.

[13]   Nii, P., "The Blackboard Model of Problem Solving and the Evolution of Blackboard Architectures: Part II," *AI Magazine*, Vol. 7, No. 3, pp. 82-106, 1986.

[14]   Philips, J. B., "Enhancing Production through Automated Tool Assembly Design," *Proceedings of the Seventh Annual International Computervision User Conference, Vol. 1*, pp. 22-34, 1985.

[15]   Sathi, A., Morton, T. E. and Roth, S. F., "Callisto: Intelligent Project Management System," *AI Magazine*, Vol. 7, No. 5, pp. 34-53, 1986.

[16]   Stiny, G., "Introduction to Shape and Shape Grammars," *Environment and Planning B*, Vol. 7, pp. 343-351, 1980.

[17]   Tomiyama, T. and Yoshikawa, H., "Requirements and Principles for Intelligent CAD Systems," in *Knowledge Engineering in Computer-Aided Design*, Gero, J. S., Ed., Elsevier Science Publishers, pp. 1-23, 1985.

[18]   Waters, R. C., "KBEmacs: Where's the AI?," *AI Magazine*, Vol. 7, No. 1, pp. 47-56, 1986.

[19]   Woods, D. D., "Cognitive Technologies: The Design of Joint Human-Machine Cognitive Systems," *AI Magazine*, Vol. 6, No. 4, pp. 86-92, 1986.

Chapter 12
DICE: AN OBJECT-ORIENTED PROGRAMMING ENVIRONMENT FOR COOPERATIVE ENGINEERING DESIGN

D. Sriram, R. D. Logcher, N. Groleau, and J. Cherneff

ABSTRACT

The development and testing of knowledge based computer tools for the integration and coordination of various phases and participants of the engineering process are described. A system architecture -- DICE -- is presented which is intended to provide cooperation and coordination among multiple designers working in separate engineering disciplines, using knowledge to estimate interface conditions between disciplines, recording who used any piece of design data created by others, and how such data was used, and checking for conflicts among disciplines, manufacturability, and manufacturing cost and schedule impacts of design decisions. The system is being developed using object-oriented programming and blackboard control techniques. Current status of DICE, along with examples in the domain of civil engineering are presented.

12.1. INTRODUCTION

On July 17, 1981, two skywalks in the lobby of the Hyatt Regency Hotel in Kansas City collapsed. It was cited as the "most devastating structural collapse ever to take place in the United States"; 114 people died and 186 were injured [27]. This was not only a failure of a physical structural system, but also a failure of the process by which most projects in the U. S. are designed and built. The primary objective of our current research is to provide computer based tools which would help during design and construction to avoid errors of the type made in Kansas City.

Artificial Intelligence in Engineering Design
Volume III

303

The Hyatt failure was attributed to a combination of three events. First, in progressing from the preliminary to detailed design, where joint and connection detailing occurs, the design of the hanger to spandrel beam connection was inadequate. Second, in developing shop drawings, the connection detail was changed by the steel fabricator, thereby "compounding an already critical condition." Third, this second error was not caught during approval checking of the shop drawings by the structural engineers. These were all errors of communication and coordination in the design process, errors caused by the structure of the process, lack of tools used in this process, and focus on documenting the product of design while neglecting "process" and "intent" documentation. These problems also exist in other engineering application areas.

Large engineering projects involve a large number of components and the interaction of multiple technologies. The components included in the product are decided in an iterative design process. In each iteration, interfaces and interface conditions among these components are designed with slack to account for potential variations created when the components and interface values become better known. Iteration proceeds towards increasing detail; design personnel may change, and their numbers expand with increasing level of detail.

The problems facing the engineering industry in the U. S. will be highlighted by considering the design and construction[2] of structures. On a single project, interacting design technologies often come from separate firms or functional groups within a firm, and there is little coordination between designers and contractor(s) during design. Because designers find coordination among themselves difficult, they leave this task to construction managers or the contractor. Thus, working drawings, used to inform the contractor of the product, lack detail. Shop or fabrication drawings are required from the contractor to document details, but potential conflicts among trades are often unrecognized until construction begins. Several undesirable effects are caused by this lack of coordination.

1. The construction process is slowed, work stops when a conflict is found.

2. Prefabrication opportunities are limited, because details must remain flexible.

3. Opportunities for automation are limited, because capital intensive

---

[2]Manufacturing in the civil engineering industry is known as *construction*. There are several differences between the construction industry and the manufacturing industry. For example, in manufacturing several hundreds of a single type of product are produced, whereas construction involves the production of one-of-a-kind products. However, the overall engineering process is similar. In this paper, the terms "manufacturing" and "construction" will be used to denote the *realization or creation of a designed artifact*.

high speed equipment is incompatible with work interruptions from field recognized conflicts.

4. Rework is rampant, because field recognized conflicts often require design and field changes.

5. Conservatism prevades design, because designers provide excessive slack in component interfaces to avoid conflict.

6. The industry is unprepared for the advent of automated construction, as the need for experience in design limits choice to available materials placed by hand.

All of these problems decrease productivity. In addition, failures, such as the Hyatt collapse, occur more often then they should. Overcoming these problems requires significant changes to the design process, together with superior *computer integrated design and construction/manufacturing* (CIDCAM) tools. Those tools must be tailored to the needs of designers who are "constantly engaged in searching out various consequences of design decisions [especially those made by others]" [2].

This paper details the development of a prototype system to test new concepts for computer tools to integrate various stages involved in the engineering of a product. The major objectives of this system are to:

1. Facilitate effective coordination and communication in various disciplines involved in engineering.

2. Capture the process by which individual designers make decisions, that is, what information was used, how it was used and what did it create.

3. Forecast the impact of design decisions on manufacturing or construction.

4. Provide designers interactively with detailed manufacturing process or construction planning.

5. Develop intelligent interfaces for automation.

Computer aided tools, which will be collectively called DICE (Distributed and Integrated environment for Computer-aided Engineering), are being currently developed to address the above objectives. DICE will significantly improve productivity by:[3]

---

[3]Engineers from several industries that we visited felt that computer aided tools for cooperation and coordination can greatly increase their productivity.

- reducing error in design;

- providing more detailed design;

- providing better manufacturing or construction planning;

- allowing easier recognition of design and manufacturing (construction) problems;

- using manufacturability criteria throughout design; and

- advancing automation.

Lessons from the Hyatt failure show that such tools are required. Had the connection designer had access to the concepts of load transmission underlying the preliminary design, local buckling might have been recognized and the joint details changed. Had the fabricator preparing the shop drawings had access to that information, he would have seen that his change violated the purpose of the connection scheme. Had the shop drawing checker seen all these changes together with their intent, he would have recognized the faults in the design.

The engineering design process and problems associated with this process are described in Section 12.2. Using this background, an overview of DICE is given. Background material on computer-based technologies used in this work and systems relevant to the current work are presented in Sections 12.3. A system architecture which utilizes concepts from knowledge-based systems and database management systems is described in Section 12.4. This is followed by a description of the current status of DICE.

12.2. SCOPE OF WORK

The problems that engineers normally solve fall along the *derivation-formation* spectrum [1]. In derivation-type problems , solutions consist of identifying an outcome or hypothesis from a finite set of outcomes known to the problem solver. By contrast, in formation-type problems, the problem solver has only the knowledge of how to form the solution. A variety of problem solving techniques are invoked to arrive at a solution.

Design and manufacturing (or construction planning) problems fall at the formation end of this spectrum. Design and manufacturing are accomplished by a team of engineers, each knowledgeable in a particular aspect of the problem, but with little knowledge of the decision processes of others. Each could be considered as one of many sources of knowledge, and hence, design and manufacturing (construction) could be viewed as *a process of constructing an artifact which satisfies constraints from many sources by using knowledge which also comes from many sources.* The extent of interactions can be seen by looking at the diverse set tasks, listed in Table 12-1, that must be performed by a diverse set of professionals during the design, for example, for a high rise building [33].

Table 12-1: Tasks involved in the Design of a Tall Building

Planning	Architectural design
Spatial layout	Site planning
Preliminary structural design	Analysis modeling
Component design	Geometric modeling
Substructure design	Cost estimating
Electrical distribution design	Electrical distribution analysis
Mechanical design	Mechanical analysis
HVAC design	HVAC analysis
Vertical transportation design	Regulatory compliance
Various design critics	Fire safety analysis

As CAD/CAE becomes more widespread, each of these consultants will be performing increased amount of their work with computer tools, tools which will embody and use their knowledge in their speciality area.

From this view, the stages of design and construction might be described as:[4]

1. **Problem Identification.** The problem, necessary resources, target technology, etc., are identified at this stage.

2. **Specification Generation.** Design requirements and performance specifications are listed.

3. **Concept Generation.** The selection or synthesis of potential design solutions, such as a structural system, is performed. Several alternative designs may be generated.

4. **Analysis.** The response of the system to external effects is determined by using the appropriate model for the system.

5. **Evaluation.** Solutions generated during the Concept Generation stage are evaluated for consistency with respect to the specifications. If several designs are feasible then (normally) an appropriate evaluation function is used to determine the best possible design to refine further.

---

[4]Similar stages occur in the manufacturing industry.

6. **Detailed Design.** Various components of the system are refined so that all applicable constraints (or specifications) are satisfied.

7. **Design Review.** The detailed design is checked for global consistency.

8. **Construction.** This involves the preparation of shop drawings, development of detailed construction schedules, actual construction, and construction monitoring.

There may be significant deviations between the properties of components assumed or generated at the Concept Generation stage and those determined at the Detailed Design stage, which would necessitate a re-analysis. The process continues until a satisfactory or optimal design is obtained.

During each stage in this process, representatives from the various interacting disciplines meet and discuss potential interactions between the components they envision designing. They use estimates of space needs, structural, heat, and electrical loads, and other factors to set requirements for their systems based on the needs of others. Experience is used to estimate these interfaces. Explanations on how these estimates were determined is seldom sought, except where they cause conflicts between objectives. When individual designers select components and systems during any stage of design, they use and try to develop solutions which satisfy the interface estimates.

The problem with this process is that individual designers often lack sufficient experience in both estimating their interfaces (assessing their impact on others) and in asking for information needed from others. They assume, instead, situations similar to other designs. Similarly, they seldom think about and may even lack knowledge of constructability or management and control of the construction process. This may lead to incompatible component selection and poor choice of design parameters. For example, the use of wide rooms in low cost housing is incompatible with inexpensive construction techniques. The designer is assumed in this process to have sufficient knowledge of construction techniques, materials, and equipment to make proper decisions. This is seldom the case. Also, since the present design process does not document reasons behind decisions, others cannot easily question decisions or improve designs. DICE's framework was developed to obviate the above problems. A simplified view of DICE is shown in Figure 12-1, where users within their discipline interact with individual CAD tools and KBS for component design and solution generation. These systems automatically communicate with a global system which provides data and support facilities.

An initial (prototype) version of DICE operates on two SUN workstations. The final system will operate on several interacting computers, which will be high speed workstations with good knowledge representation tools. Knowledge representations for support facilities are required to:

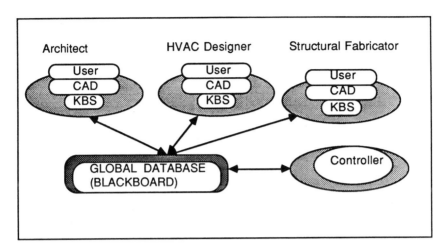

Figure 12-1: A Simplified View of a Distributed Design Environment

1. Estimate and negotiate interface parameters between stages of design, doing so in an interactive manner, when a designer asks for information (i.e., if a designer asks for information that has not yet been developed, knowledge will be used to estimate values);

2. Keep track of who used design information, when, and whether it was estimated or actual values (so that when values change, the design can remain coordinated);

3. Use coded individual knowledge sources to assist in or automate component design, retaining component information about sources of data used in the design, the algorithms or knowledge used, and inputs on design rationale from the user;

4. Operate numerous background processes to check design choices for interferences, violations of interface assumptions, constructability, and cost and schedule impacts;

5. Allow user input and design alterations from either a commercial CAD system or the knowledge representation workstation; and

6. Inform designers of the impacts of initial designs and changes by others on their design choices.

We extend the simplified version of DICE, shown in Figure 12-1, to address the above issues in Section 12.4.

12.3. BACKGROUND

There are five technologies are required to realize DICE: Artificial Intelligence (knowledge-based systems, object-oriented programming, negotiation theory, etc.), Distributed Databases, Local Area Networks, Design Methodologies (design for assembly, Taguchi's methods, house of quality, hierarchical design models), and Visual Computing (geometric reasoning and user interfaces). In the next section, we will describe some of the computer-based technologies that are utilized in the current DICE prototype. This is followed by a summary of work on negotiation. A review of relevant research work is provided in Section 12.3.3.

12.3.1. Relevant Computer-Based Technologies

Developments in computer science and engineering methodologies have provided engineers with a variety of software development tools. The computer-based software development tools that are relevant to this project are:

1. Object-Oriented Programming (OOP) Methodologies;
2. Knowledge-based systems (KBS);
3. Database management systems (DBMS);
4. Visual computing; and
5. Local area networks.

Object-Oriented Programming. Object-Oriented Programming (OOP) is a style of programming that involves the use of objects and messages. Objects are defined by Stefik and Bobrow [40]:

Objects are entities that combine the properties of procedures and data since they perform computations and save local state.

Objects contain slots and slots may consist of a number of facets. A slot may simply be an attribute or it may be a relation. The facets contain meta-information about the slot.

All actions in object-oriented programming are performed through messages. Messages tell the object *what to do* and not *how to do it*. Methods are attached to the object to execute the actions associated with the messages. The message-passing ability in OOP supports the concept of *data abstraction*.

Objects can be grouped into "classes," where each class of objects knows how to perform several actions. Individual instances of objects can be created from a particular class. The Object-Oriented programmer builds a system by specifying new classes of objects and their associated methods. Most OOP systems support the concept of "inheritance," where a class of objects may be specified as a "subclass" of another "superclass" of objects. Subclasses and instances inherit methods from their superclass, and are usually more specific entities than their (usually) more general superclass. An object may inherit methods and data from multiple classes through a network of structural relationships. In short, every object has the ability to: *store information, process information, create new information*, and *communicate with other objects. Thus OOP facilitates encoding design and construction knowledge in a disaggregated and modular form.*

As an example consider the following object:

```
BEAM-1
   instance: "Beam"
   M :
   Methods: Display-moment, Calculate-moment
```

The message *(send beam-1 calculate-moment)*, where beam-1 is the object to which the message is addressed, would compute the moment in accordance with the Calculate-moment method. For further details about object-oriented programming see [40].

Knowledge-based systems. Knowledge-based systems (KBSs) are computer programs which incorporate knowledge and reason through the application of their knowledge to data about a specific problem. If these systems incorporate human expertise then they are called knowledge-based expert systems (KBES)[5]. A typical KBES consists of three components: *Knowledge-base, Context*, and *Inference Mechanism or Control Mechanism.* Several problem-solving architectures used in the Inference Mechanism are described in [39].

In this work, a knowledge-based framework - the Blackboard architecture - that facilitates the integration of diverse sources of knowledge is used [17]. In addition, the work on truth maintenance systems will also be utilized [10, 11].

The Blackboard architecture provides a framework for: 1) integrating knowledge from several sources, and 2) representing multiple levels of problem decomposition. It uses two basic strategies [28]: 1) divide and conquer, and 2) opportunistic problem solving. The divide and conquer strategy is realized by decomposing the context, which is called a *Blackboard*, into several levels depicting the problem solution decomposition, while opportunistic problem solving is achieved by focusing on the parts of the problem that seem promising.

---

[5]For the purpose of this paper, the term KBS and KBES will be used interchangeably.

The Blackboard architecture has been successfully used in solving a wide range of tasks, such as speech recognition [13], signal processing [29], and planning [17].

Database management systems. Engineers have always dealt with large amounts of data in diverse applications. Hence, storing and manipulating data forms an integral part of the engineering process. Database management systems (DBMS) provide means to store large amounts of data in databases for use by a variety of applications. Data access is controlled through a dictionary so that individual programs need not be changed when the database structure changes. If a problem requires the integration of several geographically distributed databases then we enter the realm of distributed databases. A system that manages these distributed databases is termed as a distributed database management system (DDMS). There are several issues that arise in the development of a DDMS: concurrency control, query processing, reliability, efficiency, etc. (See [3, 7, 30] for further discussion of these issues.)

Visual Computing. Engineers make extensive use of diagrams (images) to convey their ideas. They also like to see scientific information (or data) to be conveyed by visual diagrams (images). Hence computer-based systems for engineering should have the ability to: 1) recognize and understand diagrams, and 2) generate diagrams. The study and development of the methodologies required to provide above capabilities in a computer program falls under the realm of *Visual Languages*. Visual languages can be classified into: 1) Visual Information Processing Languages (VIPL), and 2) Visual Programming Languages (VPL) [4]. In VIPL, the objects that are displayed on the screen (by the engineer) have inherent visual meaning, i.e., the object has some semantic meaning associated with it. The task here is to map these objects into their semantic content. An example of VIPL is spatial reasoning about engineering objects. On the other hand in VPL, the visual diagrams are generated on the screen from scientific (or otherwise) data and it is left to the engineer to extract the semantic meaning of these diagrams, e.g., current work on solid modeling. It is also important to realize that these visual languages should be portable. Hence, they should be developed in an environment that is portable across a variety of hardware, such as the X Window system, which is rapidly gaining acceptance as an industry standard.

12.3.2. Negotiation Theory

Negotiation is a process by which a joint decision is made by two or more parties. The parties first verbalize contradictory demands and then move toward agreement by a process of concession making or search for new alternatives. The parties can range in size and importance from children trying to decide how to divide up a set of toys to nations trying to end a war (which might not be as different as it seems once you really think about it). Irrespective of the size and the type of the parties involved, there seems to be a general body of principles that are applicable to address the negotiation problem. Pruitt's approaches to the negotiation problem has considerable relevancy for DICE [31].

Pruitt's Principles of Negotiation. Two basic principles of negotiation - the goal/expectation hypothesis and the strategic choice model - are presented in [31].

1. The *goal/expectation hypothesis* states that for most forms of coordinative behavior to be enacted, it is necessary to have both a goal of achieving coordination and some degree of trust in the other party's readiness for coordination.
2. The *strategic choice model* postulates three basic strategies for moving toward agreement: unilateral concession; competitive behavior; and coordinative behavior. These are assumed to be at least partially mutually exclusive. Hence, tradeoffs among them can be expected. Conditions that discourage the use of one strategy should encourage the use of the others and vice versa.

Since negotiation itself can be viewed as a form of coordination, the theory of how coordination develops provides insight into the conditions under which negotiation begins.

Coordination. Coordination occurs when bargainers work together in search of a mutually acceptable agreement. Without the possibility of coordination, negotiation would often take an inordinate amount of time, create much psychological strain, end in disagreement, and/or poison future relations between the participants. Coordination is very common, especially in the later stages of negotiation, when competitive behavior no longer seems very productive. According to Pruitt, two main types of coordination occur in negotiation [31]:

1. *Concession exchange*, in which the parties move toward one

another on a single dimension or swap concessions on different dimensions; this is a form of *compromise bargaining*.

2. *Problem-solving discussions*, in which the parties share information about goals and priorities in search of an option that will satisfy both parties' needs, that is, an *integrative agreement*.

For further background on negotiation theory see [16].

12.3.3. Relevant Systems

The following systems, developed in recent years, address some of the capabilities needed for DICE.[6] However, these systems do not address the problem from a global perspective. Further, full-scale implementations of many of these systems do not yet exist.

12.3.3.1. CAE systems

The following systems have been developed for computer aided engineering.

The DARPA Initiative in Concurrent Engineering (DICE). The primary goal of DICE2[7] is the development of an advanced design and engineering environment that facilitates rapid prototyping of electro-mechanical parts and high-density electronic assemblies [32]. In DICE2, each workstation is connected to the DICE2 Communication Channel (DCC) through which all information flows, with the help of a Concurrency Manager (CM). Each workstation has a Local Concurrency Manager (LCM). LCM coordinates the flow of Info-Paks (extended NFS-Network File Servers) with the CM to provide transparency. In addition, the workstation also has a local database (LDB) and a set of analysis and support tools, with isolated interfaces. Using, the LCM, LDB and the tools, a domain expert will be able to access remote data, perform com-

---

[6]In addition, a recent summary of some of the work in computer aided cooperative work appeared in [22]. The COLAB work at Xerox PARC and XEROX Webster is relevant to the overall scope of DICE. The work reported in [23] is relevant to the communication aspects of DICE.

[7]For the sake of preventing confusion, this system will be referred to as DICE2 and not as DICE.

putations and communicate the results to others in the network. The LCM will also be used to access External Networks and remote Compute Servers.

Conclusions drawn from local analysis are used to influence design by asserting decision parameters on a Blackboard. A Knowledge Server (KS) is used to retrieve generic information such as that which may be found in a handbook, or transformed information from the Part-Process Organization (PPO) database. The simulated Production Facility (SPF) at one end of the DCC is a multi-level simulator that may be used by production engineers, plant designers and others to validate producibility before designs are finalized. The Advanced Prototyping center (APC), at the other end of the DCC, consists of programmable machine tools which may be used to produce actual prototypes prior to production runs at the pilot plant level.

DICE2 is still in a development stage and is yet to be demonstrated on a real world problem. If successfully implemented, this system should provide answers to most of the problems stated here. It is interesting to note that DICE and DICE2 have many similar characteristics.

NEXT-CUT: A Second Generation Framework for Concurrent Engineering. Next-Cut consists of a centralized knowledge store, with different agents communicating through this central knowledge medium. The knowledge store embodies both product and process models, depicting the life cylce of the product [8]. Any changes made to this central knowledge store are communicated to the appropriate agents. Although the agents and the central knowledge store reside on a single machine, the architecture is easily extendable to a distributed environment. Next-Cut is implemented in SPE, a LISP-based object-oriented programming environment.

An Integrated Software Environment for Building Design and Construction (IBDE). Fenves *et al.* developed an integrated environment - called IBDE (Integrated Building Design Environment) - of processes and information flows for the vertical integration of architectural design, structural design and analysis and construction planning [14]. The integrated environment makes use of a number of AI techniques. The processes are implemented as KBES. A Blackboard architecture is used to coordinate communication between processes. The global information shared among the processes is hierarchically organized in an Object-Oriented Programming language.

The Integrated Building Design Environment (IBDE) system is implemented in the form of five vertically integrated Knowledge-Based processes:

1. An architectural planner (ARCHPLAN).
2. A preliminary structural designer (HI-RISE).
3. A component designer (SPEX).

4. A foundation designer (FOOTER).
5. A construction planner (CONSTRUCTION PLANEX).

The processes communicate with each other in two ways:

1. a message Blackboard is used to communicate project status information such as whether a process is ready to execute, has successfully performed its task or has encountered a failure, and

2. a project database used for storing the information generated and used by the processes

A controller uses the information posted on the Blackboard to initiate the execution of individual processes. The controller also directs the data manager to provide and receive the information shared between the processes. Since the different processes may reside on different machines, the data manager and the Blackboard rely on a local area communication network.

DESTINY. DESTINY is a knowledge-based framework developed for integrating all stages of the structural design process [37]. It consists of several Knowledge Modules (KMs) that communicate through a Blackboard. The Blackboard consists of several levels that define the abstraction hierarchy of objects. The entities generated at various levels in the Blackboard are connected through relational links to form a solution to the structural design problem. The KMs are grouped into: *Strategy, Specialist*, and *Resource* KMs. The Specialist KMs perform various tasks of the structural design process. Example KMs are: ALL-RISE, for preliminary structural design; MASON, for structural analysis; DATON, for detailing; and CRITIC for evaluating designs. The Strategy KM controls the design process, while the Resource KMs contain algorithm programs such as Finite Element Analysis.

KADBASE. KADBASE was developed to provide a knowledge-based interface for communication between multiple knowledge-based expert systems and databases [18]. The main components of KADBASE are: 1) The Knowledge-based System Interface (KBSI), which provides the translations (semantic and syntactic) for each KBES communicating with the Network Data Access Manager (NDAM); 2) Knowledge-Based Database Interface (KBDI), which provides the translations needed for each DBMS for communicating with NDAM; and 3) NADM, which decomposes queries and updates and sends them to the appropriate KBES/DB. A good review of of the work on the use of databases in engineering can be found in [18].

GEMSTONE, ORION and ENCORE. GEMSTONE combines the advantages of Object-Oriented Programming (OOP) languages with the storage management of traditional DBMS [26]. In addition to the OOP language (called OPAL), GEMSTONE's programming environment also provides an interactive interface: for defining new objects; 2) a windowing package for building user interfaces; and 3) interfaces to conventional programming languages. GEMSTONE was developed for a multi-user environment and has a disk manager for swapping objects to and fro from resident memory into the hard disk. ORION and ENCORE are object-oriented DBMS developed for computer-aided engineering at M.C.C. and Brown University, respectively; ORION (currently marketed as ITASCA) is a LISP-based system, while ENCORE is a C-based system (Details GEMSTONE, ORION and ENCORE can be found in [25]).

Katz's Work. Several issues in the management of large design databases, in the realm of VLSI design, are discussed in [19]. The design data management system, described in [19], consists of the following components: 1) Storage component, which manages design data on secondary storage; 2) An object-oriented database, which supports several semantic relationships between objects of the database; 3) Design Librarian, which coordinates all access to shared design data by making design objects available to various workstations; 4) Recovery subsystem, which manages changes from workstations to database servers; 5) Design Validation subsystem, which assists in determining the validity of an object when changes occur; and 6) Design Transaction manager, which uses the Design Librarian, Recovery subsystem, and the Design Validation subsystem, to ensure that the design objects created are in a consistent state.

Eastman's Work. Eastman and his colleagues have addressed the issue of integrating multiple design databases in considerable detail [12]. Eastman points out, with an example of piping system design, that a DBMS must be able to handle interactions of several forms, such as: "the loads generated by piping that must be picked up by the structure" and "spatial conflicts between piping and structural systems." He recommends the use of transaction graphs, where nodes denote transactions on a database and links provide the precedence relationships, as a potential solution to the above problem. In addition, he also addresses the problem of concurrent users and communication among multiple users. He proposes that a design database should "have attached methods that other users of the database may be made aware of any assumptions." Although Eastman addressed several important issues in his work, a full scale implementation was hindered by lack of adequate programming methodologies, such as OOP systems and KBS, at that time.

12.3.3.2. Computer-aided negotiation

The following computer-based systems address the negotiation problem.

CALLISTO. The engineering of large complex artifacts involves a number of activities which require the close cooperation of a number of departments. The CALLISTO project emerged out of the realization that classical approaches to project management is inadequate [36]. A prototype, MINI-CALLISTO, was developed to support the needs of an organizational unit by providing means for communication and coordination during the pre-planning stage of the project management process. MINI-CALLISTO has several Knowledge Modules (KMs), each with a similar architecture to MINI-CALLISTO, that communicate through messages. The coordination between various KMs is achieved through a *constraint-negotiation* algorithm, specially designed to address project management problems.

Resources Reallocation Problems. A theory of negotiation to solve resource reallocation problems among multiple agents was developed and tested by Sathi [35]; the resource allocation problem deals with the optimal allocation of resources to agents. In a typical resource allocation situation, there are a set of agents, each with a set of allocated resources working against a set of activities requiring resources. A typical reallocation transaction specifies ownership exchange for one or more resources among agents. A simple transaction involves selling of a resource from one agent to another. A trade involves a two way exchange of resources between two agents. A cascade involves an open or closed chain of buy and sell among more than two agents. In his work, Sathi defines Constraint-Directed Negotiation as a set of qualitative evaluation and relaxation techniques based on human negotiation problem solving. The three constraint relaxation techniques experimented are as follows:

1. *Bridging*: A grouping of buy/sell bids or transactions in order to meet a complex constraint.

2. *Reconfiguration*: A change in the resource attribute value in order to meet the requirements of a buyer.

3. *Logrolling*: Selective constraint violation on less important constraints in order to accept a transaction which is acceptable on more important constraints.

While initial evaluations on constraints were done locally by each agent, the above relaxations were performed during a group problem-solving session, as if performed by a mediator. The automated problem solver uses a mix of local and global knowledge to facilitate cooperative reasoning where some problem solving is done by each agent and some as a group. The individual and group search processes use several aspects of constraints, such as constraint importance and looseness to prioritize the search steps.

Deals Among Rational Agents. A formal framework is presented in [34] that models communication and promises in multi-agent interactions. This framework extends previous work on cooperation without communication and shows the ability of communication to resolve conflicts among agents having disparate goals. Using a deal-making mechanism, agents are able to coordinate and cooperate more easily than in the communication-free model. In addition, there are certain types of interactions where communication facilitates mutually beneficial activity that is otherwise impossible to coordinate.

Negotiation as a Metaphor for Distributed Problem Solving. A framework, called *Contract Net*, that specifies communication and control in a distributed problem solver was developed by Davis and Smith [9]. The kinds of information which must be passed between nodes in order to obtain effective problem-solving behavior is the origin of the negotiation metaphor; task distribution is viewed as a form of contract negotiation. The use of the Contract Net framework is demonstrated in the solution of a simulated problem in area surveillance. The system is based on the assumption that three issues are central to constructing frameworks for distributed problem solving [9]:

1. The fundamental conflict between the complete knowledge needed to ensure coherence and the incomplete knowledge inherent in any distribution of problem solving effort.

2. The need for a problem solving protocol.

3. The utility of negotiation as an organizing principle.

Davis & Smith feel that negotiation is an important aspect of distributed problem solving and consists of three important components: 1) there is a two-way exchange of information, 2) each party to the negotiation evaluates the information from its own perspective, and 3) final agreement is achieved by mutual selection.

Multistage Negotiation in Distributed Planning. Multistage negotiation provides a means by which an agent can acquire enough knowledge to reason about the impact of local activity on non-local state and modify its behavior accordingly. Conry describes a multistage negotiation paradigm for planning in a distributed environment with decentralized control and limited interagent communication [6]. The application domain of interest involves the monitoring and control of a complex communications system. In this domain, planning for service restoral is performed in the context of incomplete and possibly invalid information which may be updated dynamically during the course of planning. In addition, the goal of the planning activity may not be achievable - the problem may be over constrained. Through multistage negotiation, which involves negotiating on primary goals at first and taking secondary goals into account later, a planner is able to recognize when the problem is overcontrained and to find a solution to an acceptable related problem under these conditions. A key element in this process is the ability to detect subgoal interactions in a distributed environment and reason about their impact.

Run-time Conflict Resolution in Cooperative Design. Klein and Lu have proposed a conceptual framework for conflict resolution in cooperative design [21]. The basic assumption in this approach is that the different kinds of conflicts that occur in design can be arranged into a set of abstract classes and that conflict resolution strategies can be associated with each conflict class. Klein and Lu define model-based negotiation as the incremental relaxation of constraints derived from one or more KSs involved in a conflict situation. Examples of constraint relaxation include broadening a numerical range or adding members to a constraint. The goal of this relaxation is to produce a satisfiable constraint set while minimizing the decrement in performance from the perspective of the conflicting KSs. This negotiation is called model-based because it is based on a model of how changes in the design lead to changes in performance. The result of such negotiation is a compromise that may be less than optimal locally, but removes the conflict situation so design can continue. To be able to engage in negotiation, each KS must be able to [21]: 1) relate suggested design changes to performance changes; and 2) express the performance changes using a common yardstick so that a reasoned decision can be made.

Negotiation of Conflicts Among Design Experts. Lander and Lesser propose two major types of negotiation operations, with examples from the domain of kitchen design, and describe the design and prototype implementation framework for knowledge-based systems [24]. They identify two types of negotiation based on Pruitt's work:

1. *Compromise bargaining.* It is appropriate when differences in proposed solutions are not too severe, constraints have some built-in flexibility and there are well-defined mechanisms for relaxing or strengthening constraints as necessary.

2. *Integrative bargaining.* When the above conditions don't apply, it may still be possible to come to an agreement by reevaluating the

long-term goals and knowledge available. There may be an acceptable solution that is not obvious in the current problem formulation but which could be discovered by restructuring some underlying assumptions.

Table 12-2 summarizes the problems addressed, technologies used, and relevancy to the current work.

12.4. A FRAMEWORK FOR A DISTRIBUTED AND INTEGRATED ENVIRONMENT FOR COMPUTER-AIDED ENGINEERING

In Section 12.1 (page 305), several objectives for a Distributed Integrated environment for Computer-aided Engineering (DICE) were enumerated. To achieve these goals, a system architecture based on current trends in programming methodologies, object-oriented databases, and knowledge based systems will now be proposed. An overview of DICE is provided in Section 12.4.1. This is followed by a discussion of various components comprising the system.

12.4.1. Overview of DICE

DICE can be envisioned as a network of computers and users, where the communication and coordination is achieved, through a global database, by a control mechanism. DICE consists of several Knowledge Modules, a Blackboard, and a Control Mechanism. These terms are clarified below.

1. **Control Mechanism.** The communication, coordination, data transfer, and all other functions define the Control Mechanism; the Control Mechanism could be viewed as an Inference Mechanism.

2. **Blackboard.** The Blackboard is the medium through which all communication takes place. The Blackboard in DICE is divided into three partitions: Solution, Negotiation, and Coordination Blackboards. The Solution Blackboard partition contains the design and construction information generated by various Knowledge Modules, most of which is referred to as the Object-Hierarchy containing information about the design product and

Table 12-2: Systems Relevant to Current Work

System	Purpose	Computer-based Technologies Used	Limitations	Relevancy to Present Work	Developer (Date)
DESTINY	An integrated KBS framework for structrual analysis and design	KBS (Blackboards), some DBMS	Only in conceptual stage. Many ideas yet to be tested.	The Blackboard framework. Concept of Strategy Knowledge Module. Symbolic and numeric coupling.	Sriram (1985, CMU)
KADBASE	Provide a KBS interface between multiple KBES and DBMS	KBS (frames) , DBMS	No coordination	Interface definitions between KMs	Howard & Rehak (1986, CMU)
GEMSTONE ORION ENCORE OTHERS	Provide an object-oriented DBMS	OOPs, DBMS	No coordination	Version control Transaction management Disk storage management	Maier, et al. (1986, Oregan) Kim, et al. (1987, MCC) Zdonik et al. (1987, Brown) See(Lochovsky 87)
Katz's Work	Provide DBMS for engineering VLSI design	Objects, DBMS	No coordination	Semantic relationships, storage management, transactions	Katz (1980-, UC Berkley)
Eastman's Work	Integrate multiple project databases	Pascal record structures, DBMS	No concrete implementation	Communications between users	Eastman (1981-84, CMU)
CALLISTO	Means for communication and coordination during preplanning stage pf Poject management process.	KBS, with OOPs flavor	Does not address interface issues	Coordination (constraint negotiation)	Sathi, et al. (1986-, CMU/CGI)
IBDE	Integrated building design	Blackboard, Uses DPSK	Simple communication, No negotiation	Same as DESTINY	Fenves, et al (1988, CMU)
DICE2	Cooperative engineering Design	Distributed Blackboard	Full implementation does not exist	Communication, Coordination, etc.	Reddy, et al. (1988, WVU)

Other Systems, as described in text

process, while the Negotiation Blackboard partition consists of the negotiation trace between various engineers taking part in the design and manufacturing (construction) process. The Coordination Blackboard partition contains the information needed for the coordination of various Knowledge Modules.

3. **Knowledge Module.** Each Knowledge Module can be viewed either as: a knowledge based expert system (KBES), developed for solving individual design and construction related tasks, or a CAD tool, such as a database structure, i.e., a specific database, an analysis program, etc., or an user of a computer, or a combination of the above. A KBES could be viewed as an aggregation of Knowledge Sources (KSs). Each KS is an independent chunk of knowledge, represented either as rules or objects. In DICE, the Knowledge Modules are grouped into the following categories: Strategy, Specialist, Critic, and Quantitative. The Strategy KMs help the Control Mechanism in the coordination and communication process. The Specialist KMs perform individual specialized tasks of the design and construction process, while the Quantitative KMs are mostly algorithmic CAD tools. The data representation (or language) used in a KM may be different from that used in the BB. Hence, each KM is provided with an interface module which translates the data from the KM to the BB and vice versa.

A conceptual view of DICE for design and construction is shown in Figure 12-2. In the DICE framework, any of the KMs can make changes or request information from the Blackboard; requests for information are logged with the objects representing the information, and changes to the Blackboard may initiate either of the two actions: finding the implications and notifying various KMs, and entering into a negotiation process, if two or more KMs suggest conflicting changes.

Details of individual components are provided in the following sections.

12.4.2. Control Mechanism

The Control Mechanism performs two tasks: 1) evaluate and propagate implications of actions taken by a particular KM; and 2) assist in the negotiation process. This control is achieved through the object oriented nature of the Blackboard and a Strategic KM. One of the major and unique differences between DICE and other Blackboard systems is that DICE's Blackboard is more than a static repository of data; DICE's Blackboard is an intelligent database, with objects responding to different types of messages.

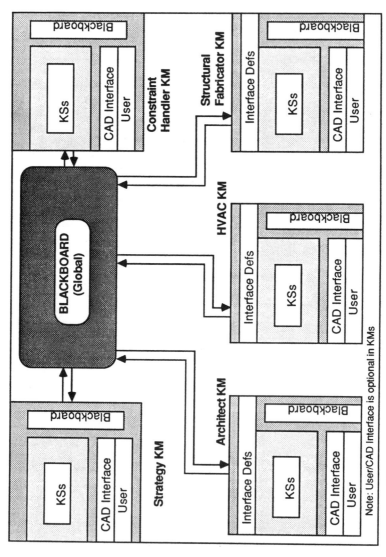

Figure 12-2: A Conceptual View of DICE for Design and Construction

Task 1 is accomplished through

1. methods associated with objects in the Object-Hierarchy of the Solution Blackboard partition (SBB); and

2. a truth maintenance system which keeps the global database in a consistent state.

If two or more KMs try to access the same object, then the priorities are determined by the Strategy KM and the scheduling information is stored in the Coordination Blackboard partition (CORDBB). A possible trace of events for the Hyatt Regency case is shown in Figure 12-3, and outlined below.

1. A preliminary design of a building (in the form of objects) which includes loading details and designer's intentions in making certain decisions is posted on to the Solution Blackboard partition by the Conceptual Designer.

2. Let the connection details of a particular joint be represented by the Connection object. The Connection Designer will send a message with details of connections and any assumptions made during the design.

3. The truth maintenance system (TMS) checks to see whether earlier assumptions made by the Conceptual Designer are violated or not.

4. Associated with the Connection object are methods, which indicate the possible KMs that can modify the object. Assume that Fabricator KM is one of them. A message is sent to Fabricator KM to find out whether the connection can be fabricated in the field.

5. Notify the Connection Designer if any problems are anticipated.

6. Sometimes two or more KMs may want to modify or access a particular object in the Solution Blackboard partition. This information is stored in Coordination Blackboard partition and is used by the Control Mechanism.

A possible scenario for task 2 for an interior design problem, which involves the cooperation of an architect and a HVAC engineer, is given below (See Figure 12-4).

1. Let the Architectural KM post the location and other details of beams in the beam object, whose primary owner is Architectural KM.

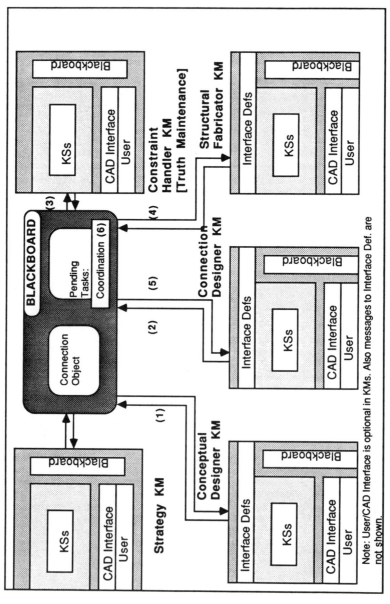

Figure 12-3: Evaluation and Propagation of Implications

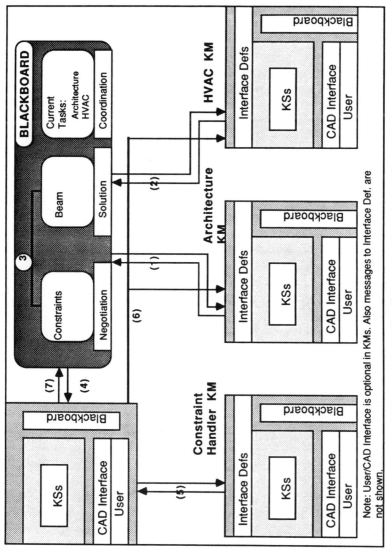

Figure 12-4: The Constraint Negotiation Process

2. The HVAC KM would post a design, which makes the assumption that ducts can pass through the beams.

3. Since the object is modified by more than one KM, Solution Blackboard partition checks to see if the object (or objects) being modified has any interaction (interface) constraints. If so then appropriate constraints are stored in the Negotiation Blackboard partition.

4. Solution Blackboard partition, then, sends a message to Strategic KM to check the constraints.

5. Strategic KM sends a message to the Constraint Handling KM (CHKM). CHKM checks to see if the interaction (interface) constraints are satisfied. If so, a message is sent to the Solution Blackboard partition and appropriate actions are taken (step 6).

6. If the interaction constraints are not satisfied then the Strategic KM performs a constraint negotiation. Constraint negotiation may involve relaxing constraints by a particular KM. If constraint negotiation fails then system goes into a deadlock and alerts the KMs. Constraint negotiation can be performed at several levels. In the current system it will be assumed that refinement of levels in the Solution Blackboard partition occurs only after appropriate interaction (interface) constraints are satisfied.

7. If above process succeeds then Strategic KM sends a message to the Solution Blackboard partition, at which stage the details required for the next level in the Solution Blackboard partition are set up and appropriate KMs are activated.

12.4.3. Blackboard: Global Database

The purpose of the Blackboard is to: 1) provide a means for storing information that is common to more than one KM; 2) facilitate communication and coordination; and 3) ensure that designs and plans generated during design and construction are consistent.

The Blackboard in DICE is partitioned into: Coordination (COORDBB), Solution (SBB) and Negotiation BB (NBB) (Figure 12-5).

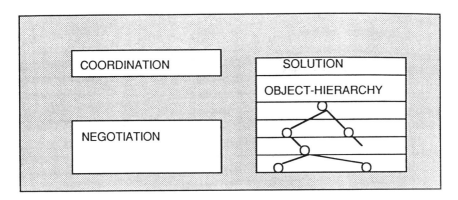

Figure 12-5: The Blackboard

12.4.3.1. Coordination blackboard partition

The Coordination Blackboard partition (COORDBB) contains the bookkeeping information needed for the coordination of KMs.

12.4.3.2. Solution blackboard partition

The Solution BB partition (SBB) is divided into levels (of the object-hierarchy). Each level contains objects that represent certain aspects of the engineering process (design and construction). The SBB does not contain all the information generated by all KMs, only information that is 1) required by more than one KM, and 2) useful in the engineering process is posted on the SBB. For example, the 3D space level will contain objects that represent spaces allocated to structural systems, piping systems, mechanical systems, etc. This level can be reduced to detailed levels, such as system and component levels.

The objects in SBB can be connected through relational links, where the relational links provide means for objects to inherit information; these relationships provide a framework to view the object from different perspectives. For the purpose of this work, the following relationships will be used in the SBB: *generalization (IS-A)* for grouping objects into classes, *classification (INSTANCE)* for defining individual elements of a class, *aggregation (PART-OF, COMPONENT)* for combining components, *alternation (IS-ALT)* for selecting between alternative concepts, and *versionization (VERSION-OF)* for representing various versions of an object. The semantics of these relationships are provided in [38]. Various planes that depict these relationships are shown in Figure 12-6.

The objects also contain justifications, assumptions, time of creation, creator, constraints, ownership KM, other concerned KMs, etc. The justification information will provide a designer's rationale and intent for the creation of the object. Assumptions made during design and construction are also stored with the object. For example, the architect, while placing the structural elements, may assume certain spatial characteristics for the HVAC systems. He may record this assumption and the rationale for such an assumption in the objects denoting the appropriate structural elements and the HVAC system. In DICE, *status* facets are associated with data attributes (slots). The *status* facet, for example, can take the following values: *unknown, assumed* and *calculated.* Additional slots will be needed for the source of data and its change, uses of data, assumptions made, etc.

Associated with these objects are methods which provide a means for: 1) performing some procedural calculations; 2) propagating implications of performing some actions, for example if the status (assumed or actual) or the value for a particular object changes then these changes can be broadcast to all concerned KMs; 3) helping to perform the coordination process. For example methods can be used as demons to perform the following construction related tasks:

1. **Estimating,** which involves continuous cost forecasting capabilities, from early estimates to detailed costs considering the equipment that will be available. This estimating will start with material and quantity modeling based on building standards for tenant work, and would first be updated with characteristics of the tenant. As layout work proceeds, material and quantity estimates would be updated.

2. **Scheduling,** which is similar in structure to Estimating, and uses much of the quantity data developed from the estimate forecast, passed to it with messages.

3. **Constructability,** where constant critics look for incompatible materials, space use, construction space needs, equipment requirements, etc.

Knowledge for all of these inputs will come from working with expert on all phases of the project, owner, designer and constructor. Further details of the use of methods in the communication process are provided in Section 12.4.2.

A typical object that resides in the SBB is structured as follows:

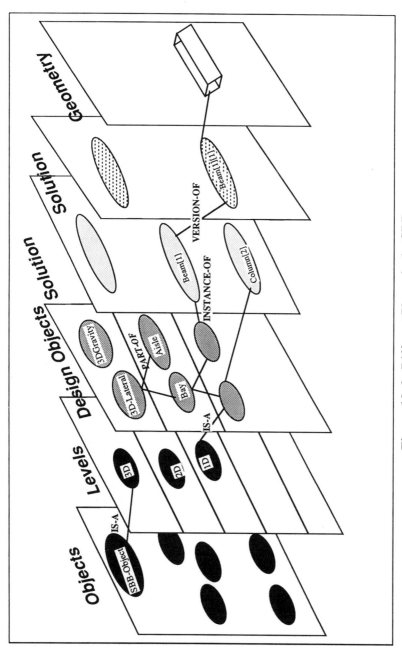

Figure 12-6: Different Planes in the SBB

```
SBB-Object
    NAME:
    VALUE:
    status:
    CREATED-BY:
    JUSTIFICATION:
    PART-OF:
    IS-ALT:
    VERSION-OF:
    VERSION-NO:
    OWNED-BY:
    CONCERNED-KMS:
    CONSTRAINTS:
        range: (IS-A CONSTRAINT-OBJECT)
        ------- (and so on)
    METHODS:
```

12.4.3.3. Negotiation blackboard partition

The Negotiation BB (NBB) could be viewed as consisting of two parts. The frst part contains various interaction constraints that are imposed on the designed object. These constraints are developed during the definition of various levels in SBB.[8] The second part consists of a trace of the negotiation process. The Negotiation BB is at an orthoganal plane to the Solution BB.

An object describing an interaction constraints is:

```
Constraint-Object
    CONSTRAINS:
            range: (IS-A SBB-OBJECT)
    INTERACTION:
            status:
            range: (IS-A INTERACTION-CONSTRAINT)
    OTHERS: (as needed)
    METHODS:
```

---

[8]Constraints in engineering design can be of two types: constraints local to the object (designed) and interaction (interface) constraints that several objects should satisfy simultaneously. An example of a local constraint is *Beam.bending-stress should be less than 0.66*Beam.material.yield-stress*, while the example of an interaction constraint is *Pipes greater than 2 inches cannot go through steel beams or columns.*

The *status* facet can take values like *satisfied, suspended, violated*, etc. A taxonomy of these constraints can be defined by the user. Adequate facilities will be provided for the user to incorporate these constraints.

12.4.4. Knowledge Modules

The Knowledge-base (KB) consists of a number of Knowledge Modules (KMs). Each of these KMs are further decomposed into small units called Knowledge Sources (KSs). The architecture of most KMs is similar to the over-all architecture of DICE, i.e., knowledge is distributed among several objects (or KSs) and communicate through message passing. KSs can also be decomposed into smaller units, if desired. Thus the KB reflects the *hierarchical design process*.

Some KMs may incorporate both *textbook* and *heuristic* (surface) knowledge, while other KMs may include *deep* knowledge. Surface knowledge consists mainly of *production rules* encoding empirical associations based on experience. This type of knowledge is useful for setting interface constraints between dis-ciplines and between levels of design interaction.   In a system with deep knowledge, both causal knowledge and analytical models would be incor-porated. A fully deep system may be difficult to realize with the current state of the art of KBES. However, it is possible to encode analytical models. In this study, the term "fairly deep knowledge" will be used to denote analytical models.

The KMs, although distributed, can be classified into the following categories: the *Strategy, Specialist,* and *Quantitative KMs*. These KMs are briefly described below.

- **Strategy KMs** analyze the current solution state to determine the course of next action. A scenario using the Strategic KM is described in Section 6.2.   Since this level may used to control various tasks, such as the activation of Specialist KMs during the coordination process, it comprises the *task control knowledge*.

- **Specialist KMs** contribute to various stages of design and construc-tion (or manufacturing).   Most KMs at this level are KBES that have a local *Blackboard* which may be divided into various levels of abstraction, and several KSs that interact with the local BB. The possible KMs that could be used in DICE for the domain of interior finishes are:

    1. *Architectural Designer*, for layout and finishes, including flooring and ceiling systems, etc.

2. *HVAC*, for heat load calculations, duct layout, diffusers, etc.

3. *Lighting*, for layout, lighting levels, heat generation, etc.

4. *Plumbing*, for layout, etc.

5. *Construction Planning*, for schedules, costs, constructability checking, etc.

6. *Structural*, only for detailing attachments.

Individual KMs will, most probably, be residing on different machines and will make extensive use of networking protocols for communicating with the Blackboard.

• **Critic KMs** contain the knowledge and mechanisms to check the consistency of the designs. Constraint Handling KM is an example of a Critic KM.

• **Quantitative KMs** contain the analytical knowledge and reference information required for analysis and design. These KMs are typically comprised of algorithmic programs and databases. Quantitative KMs comprise the algorithmic knowledge of the domain. The Specialist KMs mostly communicate with the Quantitative KMs through a Blackboard that is local to the Specialist KM.

The user forms an integral part of these KMs. An important issue in the development of KMs is the man-machine interface and how the information generated by the user is transmitted to other KSs. We assume that the user interacts with the computer through a high resolution bit mapped display (or appropriate system). Hence, there is a need to provide the appropriate semantic translations from the information provided by user to the form required by other KMs or KSs. In DICE this is achieved by the interface definition module. Further changes made by the user will be recorded in the local and global Blackboards (if needed) and appropriate actions triggered. Hence, the user can be viewed as a KS taking part in the solution process.

The KMs (mostly Specialist and Strategy) can post and retrieve information from the global Blackboard. However, an object (and associated attributes) in the Blackboard can have varied connotations (most semantic) in different KMs. Hence, there is a need to define the semantic mappings (translations) between the objects in the KMs to the objects in the Blackboard. As an example, consider the object **Beam**. In the architectural KM, the beam may be defined as follows [18]:

```
Beam
        LEFT-END-COLUMN-LINE:
        LENGTH:
        WIDTH:
        MATERIAL:
        TYPE:
        DEPTH:
        VIEW
        . . . .
```

While the same object may be defined in the HVAC KM as:

```
Beam
        L-END:
        R-END:
        D:
        TYPE:
        MATERIAL:
        R-END-MOMENT:
        . . . .
        METHODS: possible-cut-outs
```

In the Blackboard, the same object may be defined as:

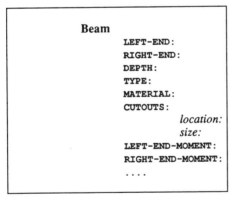

```
Beam
        LEFT-END:
        RIGHT-END:
        DEPTH:
        TYPE:
        MATERIAL:
        CUTOUTS:
                    location:
                    size:
        LEFT-END-MOMENT:
        RIGHT-END-MOMENT:
        . . . .
```

In DICE the methodology used in [18] is being adapted for developing the necessary semantic translations.

12.4.5. The Negotiation Activity

The negotiation process takes place once a conflict is detected by the Truth Maintenance System. Negotiation is achieved through the help of the Strategic KM.

Conflicts can occur either due to interface constraint violation or due to contradictory modifications of a single object. For example, a HVAC KM can decide to place pipes at the same location, where an architect had placed a beam. These conflicts can only be detected once the two designs have been posted and sufficient constraint propagation and/or modeling has been performed. Another type of constraint violation occurs when a KM changes to a partial solution posted by another KM. The two participants may or may not have similar roles in the system. For example, two architects may disagree on the location of the walkway, or the HVAC KM might want to change the depth of a beam posted by the structural engineer in order to put some pipes through it.

Once a conflict has been identified, a twofold mechanism helps the conflicting KMs in negotiating towards a mutually agreed solution. Constraint relaxation is first attempted and is followed by goal negotiation in case of failure.

The first attempt to negotiate involves traditional constraint relaxation techniques and implements compromise bargaining. Assuming that the conflict is due to constraint violations of certain design parameters, the system can act as a third party and offer compromise values to each party until an agreement is reached. In order to allow this scheme to function properly, each value posted on the Blackboard has to be accompanied by a constraint. Each constraint must specify the range of possible values. These constraints can be either *soft* or *hard* constraints.

The soft constraints can be negotiated upon and are not imperative. For example, an architect can decide that the aesthetic proportions of a particular beam imposes a certain width to depth ratio, while the structural engineer finds another ratio guided by strength requirements.

The hard constraints always have to be satisfied. These are usually bounds imposed by code regulations or physically existing constraints such as a neighboring building, etc. In this case, no relaxation is possible and another negotiation mechanism has to be triggered. In the case of hard constraints, the range of possible values is automatically set to a unique value. If the system detects a conflict involving soft constraints which have a non-empty range intersection, it can directly modify these values and notify the KM of the change. If the ranges do not show compatibility, the Strategic KM has to contact the conflicting KMs and initiate constraint relaxation. In some cases, it is hoped that the ranges themselves have soft constraints and that a compromise can be found.

The second set of techniques involves the redefinition of design goals. The KMs are asked to negotiate on a more abstract plane. It is considered that the set

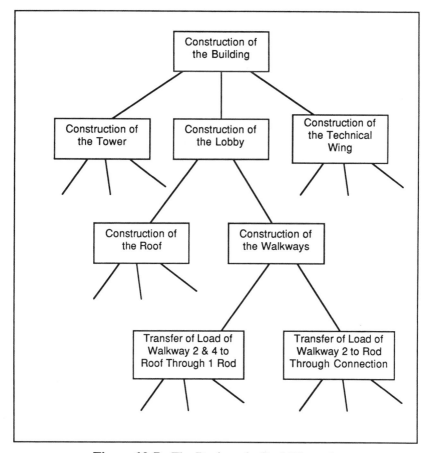

Figure 12-7: The Designer's Goal Hierarchy

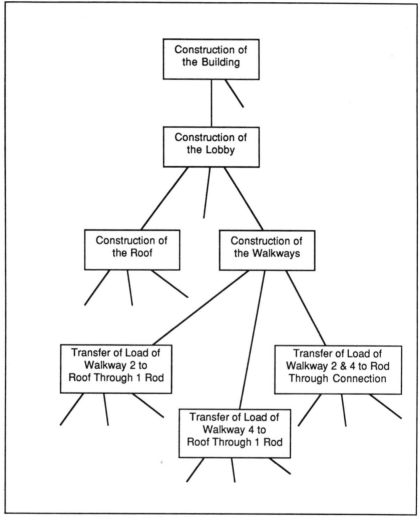

Figure 12-8: The Fabricator's Goal Hierarchy

of conflicting constraints are the concrete expression of an abstract hierarchy of goals. At the root of this hierarchy is the goal of designing and constructing the artifact for which the design team has been set up. Each participant develops his own hierarchy of personal goals (see Figure 12-7 and Figure 12-8). By helping the KMs find an agreement goal and developing a common set of more detailed goals, the system achieves integrative agreement.

The example of the Hyatt Regency walkway connection design will serve as an illustration of the multiple levels at which this goal negotiation can take place.

At the lowest level, the conflict is in the goals assigned by each participant to the connection and the rod(s). The designer wants the connection to transfer the load of only one walkway, and wants the rod to transfer the load of the two walkways to the roof. The fabricator wants each rod to transfer the load of only one walkway, and wants the connection to bear twice the load. Goal negotiation at this level can lead to a solution where two rods are connected at some level below the box-beam (see Figure 12-9). Then, each rod transfers the load of only one walkway and so does the connection.

At the next level, the system suggests that the connection need not be designed at all. That means that a different solution can be sought, avoiding the connection of the two parts of the rod and the box-beam. Such a solution could consist of hanging each walkway from its own rod directly to the roof as in Figure 12-10. The aesthetic of this solution might not please the architect, but this is another conflict story!

The next level of abstract negotiation would consist of avoiding the design of walkways. This could be done by providing fast convenient elevator service or preventing direct level to level access across the lobby, etc. Even further, the system could suggest to get rid of the lobby and design the whole building differently. By rearranging the layout, a new solution might be found which doesn't require a lobby. Finally, the participants might come to the conclusion that they have no common goal and that they do not wish to design and construct the building. Some negotiations are bound to end that way.

12.5. CURRENT STATUS

During the initial stages our major focus has been the development of: 1) utilities for defining the SBB object hierarchy, 2) transactions for posting, modifying and deleting information from the Blackboard, 3) a simulation program to demonstrate the utility of DICE, and 4) a prototype which involves the automatic generation of construction schedules from an architectural draw-

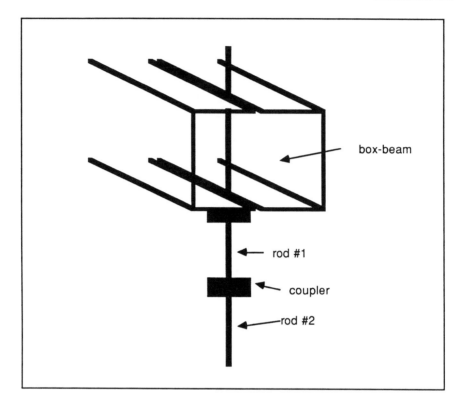

Figure 12-9: The Connected Rods Solution

ing. DICE is being implemented in a hybrid programming environment called PARMENIDES/FRULEKIT; PARMENIDES/FRULEKIT supports programming in frames and rules and was developed in LISP at Carnegie-Mellon University by Carbonell and Shell.

These topics are briefly described below.

12.5.1. Graphic Definition of Objects

The user can interactively define class objects in the Blackboard and the

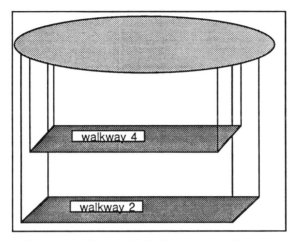

Figure 12-10: The Split Connections Solution

KMs.[9] Classes can either be created in LISP or through a menu interface provided to the user. Each class has a name, several slots which describe various attributes, and associated with each slot are facets which provide further information about the slot; the facets also contain methods.

A class object is created by clicking on CREATE-CLASS in a menu on the screen. After creating a class object, the user can display the class using the DISPLAY-CLASS option, as shown in Figure 12-11. In Figure 12-12, the class **Build1** is made a subclass of the **Hierarchy-object** class, which becomes **Build1**'s superclass. When this link is made, all slots in the **Hierarchy-object** class are inherited by **Build1**. In addition, the user can create new slots or delete slots. For example, in Figure 12-13 the user created two slots, namely NAME and HAS-PARTS. Slots can be faceted or non-faceted. The creation of facets is shown in Figure 12-14.

Instances of a class can either be defined interactively through a menu or by LISP functions. For example the function *(create-instance 'Build1 'Hyatt-regency)* would create an instance, **Hyatt-regency**, of **Build1**.

---

[9]A class denotes the grouping of objects (instance or class) which have similar characteristics, while an instance is a particular individual which belongs to a class.

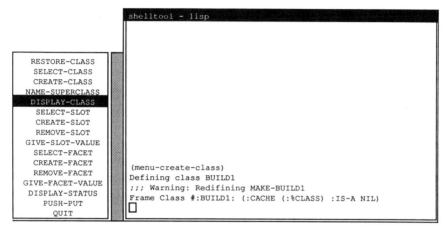

```
shelltool - lisp
```

```
RESTORE-CLASS
SELECT-CLASS
CREATE-CLASS
NAME-SUPERCLASS
DISPLAY-CLASS
SELECT-SLOT
CREATE-SLOT
REMOVE-SLOT
GIVE-SLOT-VALUE
SELECT-FACET
CREATE-FACET
REMOVE-FACET
GIVE-FACET-VALUE
DISPLAY-STATUS
PUSH-PUT
QUIT
```

```
(menu-create-class)
Defining class BUILD1
;;; Warning: Redifining MAKE-BUILD1
Frame Class #:BUILD1: (:CACHE (:%CLASS) :IS-A NIL)
□
```

Figure 12-11: Displaying a Class

```
shelltool - lisp
```

```
RESTORE-CLASS
SELECT-CLASS
CREATE-CLASS
NAME-SUPERCLASS
DISPLAY-CLASS
SELECT-SLOT
CREATE-SLOT
REMOVE-SLOT
GIVE-SLOT-VALUE
SELECT-FACET
CREATE-FACET
REMOVE-FACET
GIVE-FACET-VALUE
DISPLAY-STATUS
PUSH-PUT
QUIT
```

```
Defining class BUILD1
;;; Warning: Redifining MAKE-BUILD1
Frame Class #:BUILD1: (:CACHE (:%CLASS :LINKED-TO-ROOT :COM
PONENTS :PARENT :X :Y :SPACE :POSTED :MODIFIED :DELETED :HI
STORY :CURRENT-SOLUTION) :IS-A (HIERARCHY-OBJECT))
LINKED-TO-ROOT    (:VALUE NIL :DEPTH 2)
COMPONENTS        (:VALUE NIL :DEPTH 2)
PARENT            (:VALUE NIL :DEPTH 2)
X                 (:VALUE NIL :DEPTH 2 :CHANGEABLE T)
Y                 (:VALUE NIL :DEPTH 2 :CHANGEABLE T)
SPACE             (:VALUE NIL :DEPTH 2 :CHANGEABLE T)
POSTED            (:VALUE NIL :DEPTH 1)
MODIFIED          (:VALUE NIL :DEPTH 1)
DELETED           (:VALUE NIL :DEPTH 1)
HISTORY           (:VALUE NIL :DEPTH 1)
CURRENT-SOLUTION (:VALUE NIL :DEPTH 1 :POST-IF-SET (NOTIFY-
KMS))
□
```

Figure 12-12: Linking a Class to its Superclass

```
shelltool - /bin/csh
S))
NAME:               (:VALUE "building-1" :DEPTH 0: CHANGEABLE T)
HAS PARTS           (:VALUE NIL   :DEPTH 0)
Defining class BUILD1
;;; Warning: Redifining MAKE-BUILD1
Frame Class #:BUILD1: (:CACHE (:%CLASS :LINKED-TO-ROOT :COMPO
NENTS :PARENT :X :Y :SPACE :POSTED :MODIFIED :DELETED :HISTOR
Y :CURRENT-SOLUTION) :IS-A (HIERARCHY-OBJECT))
LINKED-TO-ROOT  (:VALUE NIL :DEPTH 2)
COMPONENTS      (:VALUE NIL :DEPTH 2)
PARENT          (:VALUE NIL :DEPTH 2)
X               (:VALUE NIL :DEPTH 2 :CHANGEABLE T)
Y               (:VALUE NIL :DEPTH 2 :CHANGEABLE T)
SPACE           (:VALUE NIL :DEPTH 2 :CHANGEABLE T)
POSTED          (:VALUE NIL :DEPTH 1)
MODIFIED        (:VALUE NIL :DEPTH 1)
DELETED         (:VALUE NIL :DEPTH 1)
HISTORY         (:VALUE NIL :DEPTH 1)
CURRENT-SOLUTION (:VALUE NIL :DEPTH 1 :POST-IF-SET (NOTIFY-KM
S))
NAME:               (:VALUE "building-1" :DEPTH 0: CHANGEABLE T)
HAS PARTS           (:VALUE NIL   :DEPTH 0)
□
```

Menu:
```
RESTORE-CLASS
SELECT-CLASS
CREATE-CLASS
NAME-SUPERCLASS
DISPLAY-CLASS
SELECT-SLOT
CREATE-SLOT
REMOVE-SLOT
GIVE-SLOT-VALUE
SELECT-FACET
CREATE-FACET
REMOVE-FACET
GIVE-FACET-VALUE
DISPLAY-STATUS
PUSH-PUT
QUIT
```

Figure 12-13: Creation of Slots

```
shelltool - /bin/csh
;;; Warning: Redifining MAKE-BUILD1
Frame Class #:BUILD1: (:CACHE (:%CLASS :LINKED-TO-ROOT :COMPO
NENTS :PARENT :X :Y :SPACE :POSTED :MODIFIED :DELETED :HISTOR
Y :CURRENT-SOLUTION) :IS-A (HIERARCHY-OBJECT))
LINKED-TO-ROOT  (:VALUE NIL :DEPTH 2)
COMPONENTS      (:VALUE NIL :DEPTH 2)
PARENT          (:VALUE NIL :DEPTH 2)
X               (:VALUE NIL :DEPTH 2 :CHANGEABLE T)
Y               (:VALUE NIL :DEPTH 2 :CHANGEABLE T)
SPACE           (:VALUE NIL :DEPTH 2 :CHANGEABLE T)
POSTED          (:VALUE NIL :DEPTH 1)
MODIFIED        (:VALUE NIL :DEPTH 1)
DELETED         (:VALUE NIL :DEPTH 1)
HISTORY         (:VALUE NIL :DEPTH 1)
CURRENT-SOLUTION (:VALUE NIL :DEPTH 1 :POST-IF-SET (NOTIFY-KM
S))
NAME:               (:VALUE "building-1" :DEPTH 0: CHANGEABLE T)
HAS PARTS           (:VALUE NIL   :DEPTH 0)

selected class : BUILD
selected slot  : NAME
selected facet : (VALUE DEPTH CHANGEABLE)
□
```

Menu:
```
RESTORE-CLASS
SELECT-CLASS
CREATE-CLASS
NAME-SUPERCLASS
DISPLAY-CLASS
SELECT-SLOT
CREATE-SLOT
REMOVE-SLOT
GIVE-SLOT-VALUE
SELECT-FACET
CREATE-FACET
REMOVE-FACET
GIVE-FACET-VALUE
DISPLAY-STATUS
PUSH-PUT
QUIT
```

Figure 12-14: Creation of Facets

In addition to defining classes and instances, the user can also display the class hierarchy, in the form of a tree. Nodes in the tree depict classes and instances. Each node is displayed in a box with the name of the class or instance. If the name does not fit in the box then it is abbreviated. The user can drag the mouse pointer over the tree to an appropriate node. This will display the full name of the node (Figure 12-15 a). If the user wants more detail about the node, he can click on the node and he will be shown the slots of the object corresponding to the node (Figure 12-15 b). Facet information for any slot can be obtained by clicking on the slot (Figure 12-15 c).

12.5.2. Blackboard Transactions

Communication between KMs is achieved through the Blackboard. The communication channels are established in special slots of the object hierarchy in the SBB. Whenever a new KM is attached to DICE, its address is placed in a special frame in the Coordination Blackboard partition.

Currently, three types of messages can be sent to the Blackboard from the KMs (and in some cases vice versa). All messages are put in a mail-box object and processed sequentially. These messages are described below.

1. **Post** allows a KM to store an object or objects at the appropriate levels in the SBB. The syntax of post is: *(Post local-object remote-object &file)*, where *local-object* is the object or pointer to a tree of objects in a KM, *remote-object* is the object/level in SBB, *file* is the name of the file that *local-object* is stored in; the & sign indicates that the file name is optional and the system creates its own name if the file name is not provided. As soon as the Blackboard receives the posted message it accesses the appropriate *file* in the KM and updates the SBB. This process is depicted in Figure 12-16.

2. **Retrieve** gets the information from the SBB to a KM. The syntax of this command is *(Retrieve remote-object &file)*. If object does not contain any information, i.e., it has a value *nil*, in the SBB then the Blackboard relays a message across the network to the appropriate KM that can provide the information; it is assumed that objects in SBB contain the names of the KMs that can update the objects. The retrieving process is depicted in Figure 12-17.

3. **Delete** provides a KM the ability to delete information on the Blackboard (SBB). The syntax of delete is *(Delete remote-object)*. Delete does not erase predefined classes, but only removes the instances. This function is currently being updated.

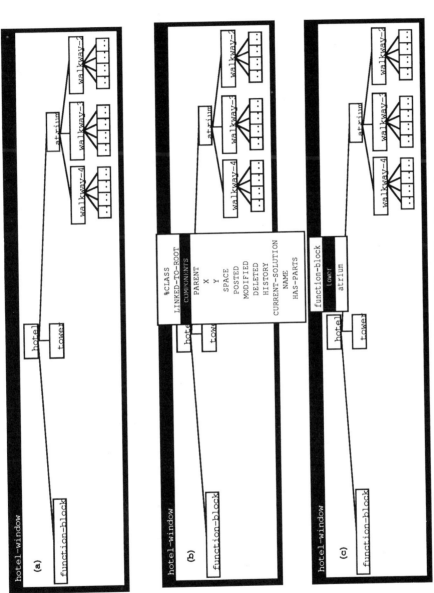

Figure 12-15: Display of the Object Hierarchy

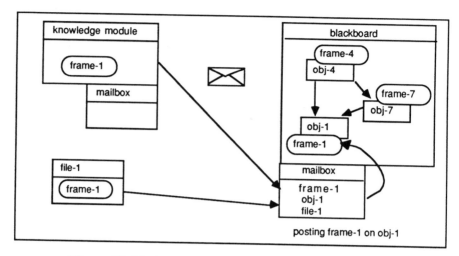

Figure 12-16: Posting Information to the Blackboard

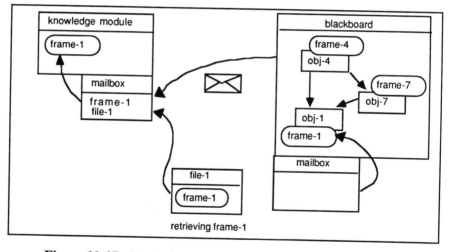

Figure 12-17: Retrieving Information From the Blackboard

12.5.3. A Simulation

A simulation of the Hyatt Regency design process was developed on two SUN computers to demonstrate some of the capabilities of DICE. A Blackboard, a Critic KM, a Constraint Manager KM, and a Strategic KM exist on the first machine (Figure 12-18), while a Connection Designer KM and a Structural Fabricator KM reside on the second machine (Figure 12-19).

The design-fabrication sequence is described below and shown in Figures 12-19 through 12-24.

1. Connection designer KM posts the connection design (denoted by 1-rod-connection) of the fourth floor walkway on the Blackboard (Figure 12-19 a).

2. Blackboard receives the design (Figure 12-20 a and b).

3. The connection object has a method that indicates that the connection design should be checked by the Critic KM. Hence, the Blackboard sends the connection design to the Critic KM (Figure 12-20 c and d).

4. The Critic KM replies that the connection design is acceptable (Figure 12-20 e).[10]

5. The Structural Fabricator KM is sent a message that a connection design has been completed and needs fabrication (Figure 12-21 a). The Fabricator retrieves the connection design, makes modifications and sends it back to the Blackboard (Figure 12-22 a, Figure 12-23 a and b).

6. Blackboard notifies the Strategic KM to check for possible conflicts (Figure 12-23 c).

7. Strategic KM retrieves the two connection (rod) designs (Figure 12-23 d) and sends it to the Constraint Manager KM to check for violation of interface constraints (Figure 12-23 e).

8. Constraint Manager KM notifies the Strategic KM that the designs are incompatible (Figure 12-23 f).

9. Strategic KM notifies this to both the Connection Designer and the Structural Fabricator (Figure 12-22 b and c, Figure 12-24 a and b).

...we assume here

---

[10]In the actual design the original connecti~
that it is an acceptable design.

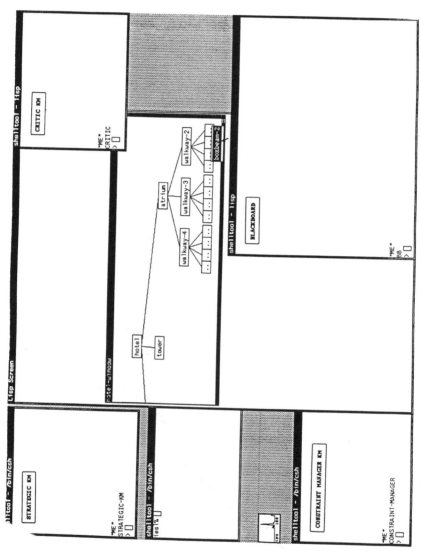

Figure 12-18: Set up for Simulating the Hyatt Regency Design Problem

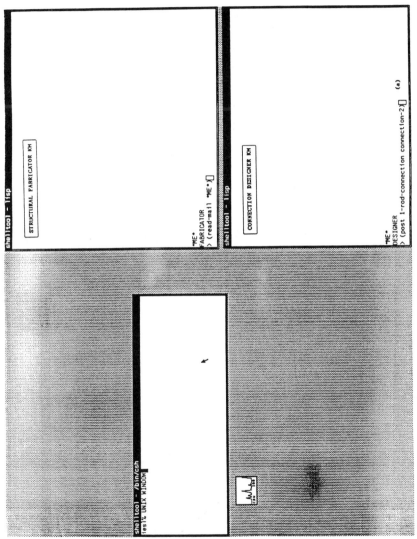

Figure 12-19: Setup for Simulating the Hyatt Regency Design Problem (Cont.)

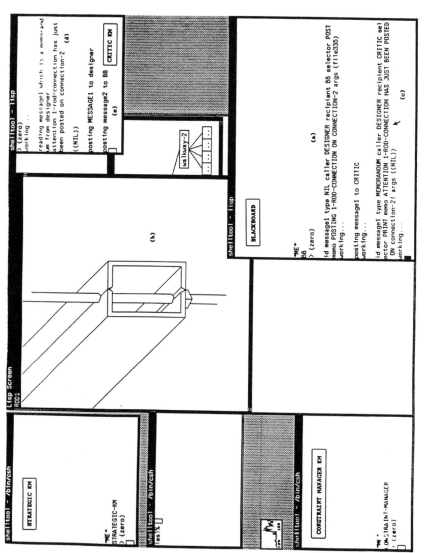

Figure 12-20: Simulation

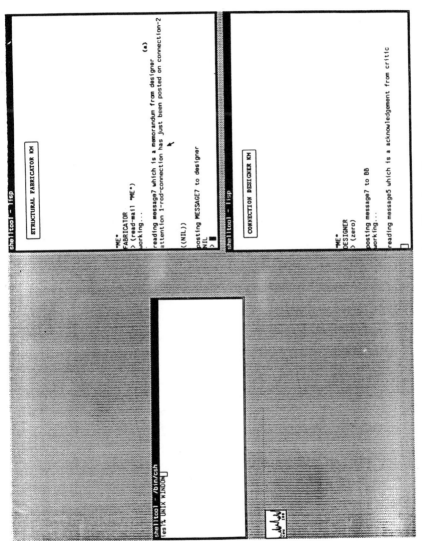

Figure 12-21: Simulation (Continued)

352 SRIRAM, ET AL.

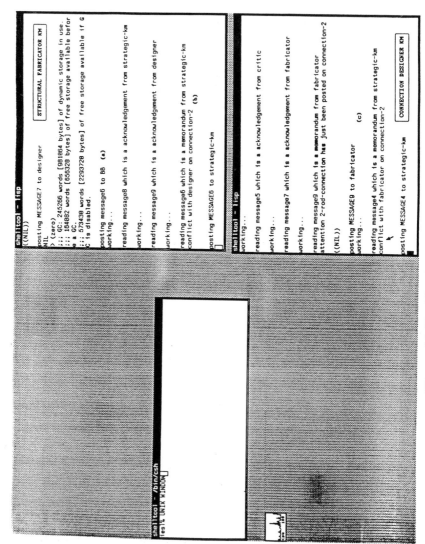

Figure 12-22: Simulation (Continued)

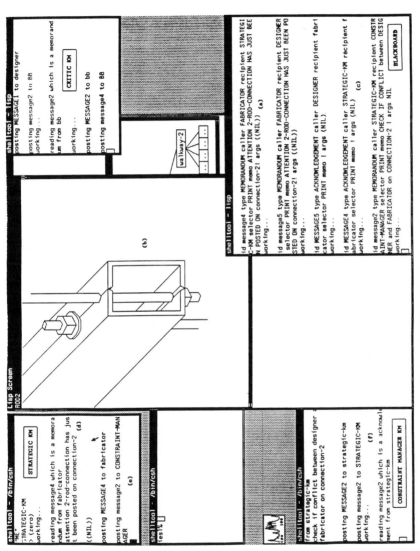

Figure 12-23: Simulation (Continued)

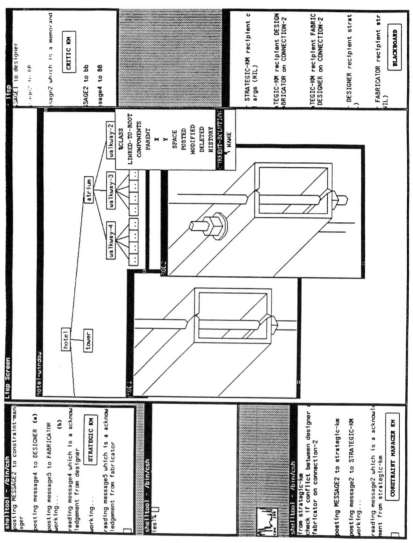

Figure 12-24: Simulation (Completed)

The above simulation program was completed in December 1987. In the next section, a project that was undertaken during January 1988 to September 1988 to demonstrate the interface definition modules is described.[11] Implementation details can be found in [16].

12.5.4. BUILDER in DICE

BUILDER automates the task of generating and maintaining schedules from architectural drawings. The initial version of BUILDER [5] was developed in KEE[TM], which is a hybrid knowledge-based programming environment [15, 20]. In this version, BUILDER had three major components - a drawing interface, a construction planning expert system, and a CPM algorithm - implemented as a layered knowledge-base, as shown in Figure 12-25. The various components of BUILDER are briefly described below.

1. **Drawing Interface.** The drawing interface layer provides for graphic input of an architectural plan. It is a menu-driven drafting system that incorporates the following features.

   a. Provides a convenient drawing system.

   b. Does the initial processing necessary to identify and classify the building components in a drawing, producing a representation of the drawing using a frame-based representation.

   c. Extracts the geometric features and produces a semantic network representation of the drawing; this semantic network representation links together the frame representation of building components.

   The friendly interface is facilitated by access to the underlying knowledge structures about building components. The menu driven system can automatically access the meanings of the symbols that it draws.

2. **Construction Planning KBES.** In an architectural drawing, the semantics of objects is normally not explicitly represented. For example there may be doors, walls, and plumbing in the drawing, but

---

[11]The NBB was not utilized in the above simulation.

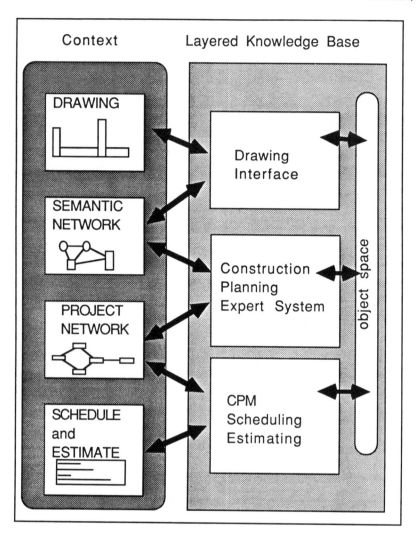

Figure 12-25: Schematic View of BUILDER

information about ordering materials for walls and doors, or having the plumbing inspected is not encoded. Neither is there any information about sequencing of tasks, or task durations, quantities, and costs. The first step in scheduling the job is to make a complete list of the tasks that need to be done. BUILDER utilizes an object-base, which is a database of engineering entities represented as frames (or objects), to complete the task list. Rules about construction methods are then activated to generate the precedence relationships between tasks. Next, BUILDER accesses a conventional database and generates an estimate of the quantities required and associated costs.

3. **CPM Algorithm.** Object-oriented and conventional CPM algorithms are implemented in BUILDER. The object-oriented approach offers some efficiency and modularity over the traditional technique in project updating, reporting, and modifying. The standard CPM algorithm is implemented for initial scheduling efficiency.

In the second version of BUILDER - DICEY-BUILDER, we are implementing the above three components as three separate KMs, as shown in Figure 12-26. The purpose here is two fold: 1) to demonstrate communication between heterogeneous KMS, and 2) to utilize this prototype to develop a protocol mechanism - similar to the Local Area Network's OSI model - for the domain of building design and construction.

The Blackboard in DICEY-BUILDER is represented as frames in PARMENIDES, while the KMs are implemented in KEE. The translation to the Blackboard from a KM and vice versa is achieved by first transforming the frames to an intermediate representation language (IRL) and then translating from IRL to the appropriate KM; the syntactic and the semantic translations are similar to the approach described in [18]. The initial Blackboard structure is generated using the editing facilities described earlier (Figure 12-27). Figure 12-28 a shows an object in the DRAW-KM. The intermediate representation format, which is a list in the current implementation, is shown in Figure 12-28 b, while the corresponding Blackboard frame is shown in Figure 12-28 c.

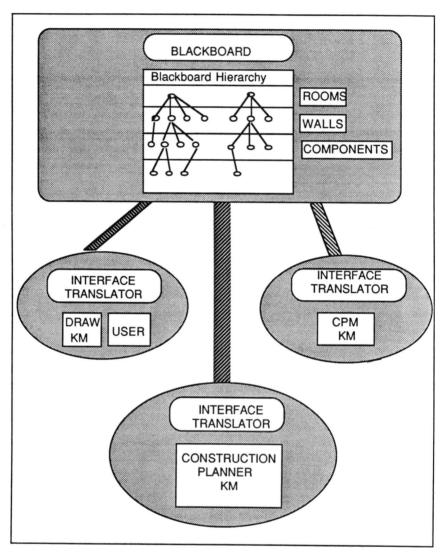

Figure 12-26: Overview of DICEY-BUILDER

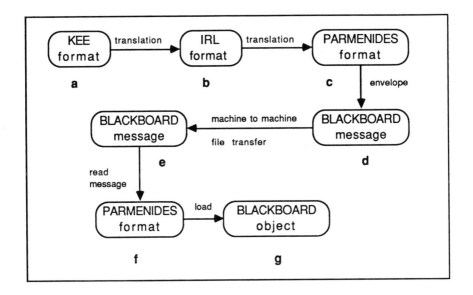

Figure 12-27: Posting From KM to Blackboard: Translation Process

12.6. SUMMARY AND FUTURE WORK

In this paper, we have described DICE, which is a collection of computer-based tools for cooperative engineering design. DICE facilitates coordination and communication in engineering design by utilizing an object-oriented Blackboard architecture, where the various participants involved in the engineering process communicate through a global database - called Blackboard. We have demonstrated the DICE framework through a simulation of the Hyatt Regency Disaster and an implementation of a construction planning KBES. Our current research addresses the following issues in the development of a distributed object-oriented CAD system.

1. **Composite Objects.** Engineering applications deal with hierarchy of objects which form a composite object. For example, beams and columns form a structural framing system. These composite objects should be treated as a single unit during storage management

Unit: **WALL144**   in knowledge base **DRAW**
Created by jonathan on 5-27-88 13:18:25
Modified by nick on 5-27-88 13:20:34
MemberOf: **WALLS**

Own slot: **BOUND**   from **WALLS**
Inheritance: **OVERRIDE.VALUES**
Values: UNKNOWN

Own slot: **CONNECTED-TO**   from **WALL144**
Inheritance: **OVERRIDE.VALUES**
Values: ((WALL185 ((311.0 95.0) (301.0 95.0))))

Own slot: **CONNECTION-MANAGER**   from **WALL144**
Inheritance: **METHOD**
Values: |DRAW>WALLS:CONNECTION-MANAGER!method|

Own slot: **CONSTRAINED-BY**   from **WALL144**
Inheritance: **OVERRIDE.VALUES**
Values: WALL185

Own slot: **CONTAINS**   from **SYMBOLS**
Inheritance: **OVERRIDE.VALUES**
Values: UNKNOWN

Own slot: **MOTION-CONSTRAINTS**   from **WALLS**
Inheritance: **METHOD**
ValueClass: **METHOD**
Values: |DRAW>WALLS:MOTION-CONSTRAINTS!method|

Own slot: **PICTURE-REPRESENTATION**   from **WALL144**   **a**
Inheritance: **OVERRIDE.VALUES**
Values: WALL.RECTANGLE145

(wall144
 (class walls)
 (contains nil)
 (picture-representation
  wall.rectangle145)
 (constrained-by wall185)
 (bound nil)
 (connected-to
  ((wall185
   ((311.0 95.0)
   (301.0 95.0)))))))

b

Frame WALL144:
 (:CACHE ())
CONTAINS NIL
PICTURE-REPRESENTATION (:VALUE WALL.RECTANGLE145)
CONSTRAINED-BY (:VALUE WALL185)
BOUND (:VALUE NIL)
CONNECTED-TO (:VALUE ((WALL185 ((311.0 95.0) (301.0 95.0)))))))   **c**

Figure 12-28: Representative Objects

and transaction processing. In DICE, a pointer to the root object is posted to the SBB. It is assumed that, during this process, other objects which are related by the SUB-PART (which is an inverse of PART-OF) link act in a passive role, i.e., the methods associated with the lower level objects are not activated. Hence, there is a need for a mechanism that allows selective activation of methods. In addition, we need to implement appropriate locking mechanisms during concurrent transaction processing of composite objects. We are planning to adapt some of the strategies utilized in several other object-oriented database systems; a survey of these systems can be found in [25].

2. **Storage Management.** In the prototype implementation of DICE, the objects in the SBB are memory resident. Since engineering design involves large data transactions, there is a need to provide effective secondary storage capabilities; by effective storage capabilities we mean that the objects should be stored such that they be retrieved efficiently. For example, all related objects can be stored contiguously on a disk. The ENCORE, OPAL, and ORION systems address this issue in detail (See [25]).

3. **Version Management.** The nature of engineering design requires that database be constantly updated. Since several participants may access and modify an object, there is a need to provide an effective version management facility. In DICE, this is achieved by the VERSION-OF relational link. Further, each object is provided with a list of KMs that can perform certain kinds of actions in the object, e.g., a connection designer KM can modify only the connection object, but it can be provided with a read access to the overall design. Our current work in this area involves adding features, such as the ability to add virtual objects which allows users to perform "what-if" simulations.

4. **Concurrency Control.** In the DICE environment, multiple engineers may try to access the SBB. This process is serialized by a locking mechanism, implemented using the UNIX^{TM} lock facility. There is a need to provide a concurrency control mechanism which allows multiple engineers to work on the same object or parts of an object-hierarchy simultaneously.

5. **Geometric Representations.** Computer aided design systems developed to date lack the flexibility to incorporate and reason about geometric entities. Since much of the communications of designers with the coordination system will be graphical in nature, the system must contain or interface with normal CAD facilities. The graphical CAD system, however, must be more intelligent and object-oriented to match the character of the blackboard (database)

in DICE. An ICAD (Intelligent CAD) graphics system is being developed. This ICAD graphics system can be used to abstract from a series of design drawings the objects existent in the design and the nature of the interfaces between them. The Blackboard should be augmented to deal with these geometric entities.

6. **X Window Interface.** The X Window system was developed at M.I.T. under the auspices of Project Athena. The primary goal of the X Window system is to provide a user interface that is portable across a variety of hardware, yet be very efficient. The version of X Windows that has been widely accepted in the industry is X.11. This will form the basis for the development of user interfaces in DICE, specially a graphical interface for defining the Blackboard objects.

7. **Constraint Negotiation.** Constraint negotiation was not implemented in the DICE prototype. We plan to extend the negotiation framework provided in this paper and provide an implementation of this extended framework.

8. **DICE for AEC.** The various levels of abstraction, objects, and appropriate KMs will be identified and a prototype will be implemented for a typical problem in the Architectural Engineering and Construction (AEC) industry. This prototype will be tested in an industrial setting.

12.7. UPDATE

This chapter was first published as a technical report in 1989. Since then considerable interest has been shown in the development of tools for computer supported cooperative product development. The following sources provide additional references:

1. Concurrent Product and Process Design, Edited by Chao, N-H and Lu, S.C-Y, Published by the American Society of Mechanical Engineers, 345 East 47th Street, N.Y. 10017, December 1989 (see also the paper by Morjaria et al.).

2. Proceedings of the Second National Symposium on Concurrent Engineering, West Virginia University, Feb., 1990.

3. Computer-Aided Cooperative Product Devlopment, Edited by Sriram, D., Logcher, R., and Fukuda, S., Springer Verlag, 1991.

12.8. ACKNOWLEDGMENTS

This research was funded by U.S. Army Research Office under contract/grant number DAAL0386G0197.

The views and/or findings contained in this report are those of the authors and should not be construed as an official department of the Army position policy, or decision, unless so designated by other documentation.

12.9. REFERENCES

[1]   Amarel, S., "Basic Themes and Problems in Current AI Research," *Proceedings of the Fourth Annual AIM Workshop*, Ceilsielske, V. B., Ed., Rutgers University, pp. 28-46, June 1978.

[2]   Barton, P. K., *Building Services Integration, E. F. N. Spon*, 733 Third Ave., NY 10017, 1983.

[3]   Cellary, W., Gelenbe, E., and Morzy, T., *Concurrency Control in Distributed Database Systems*, Elsevier Science Publishers B.V., 52 Vanderbilt Avenue, NY 10017, 1988.

[4]   Chang, S-K., Khikawa, T., and Ligomenides, P. A., Eds., *Visual Languages*, Plenum Press, New York, 1986.

[5]   Cherneff, J., *Automatic Generation of Construction Schedules from Arcitectural Drawings*, unpublished Master's Thesis, Dept. of Civil Engineering, M.I.T., Cambridge, MA 02139, 1988.

[6]   Conry, S., Meyer, R., Lesser, V., "Multistage Negotiation in Distributed Planning," *Readings in Distributed Artificial intelligence*, pp. pg 367-384, 1988.

[7]   Coulouris, G. and Dollimore, J., *Distributed Systems: Concepts and Design*, Addison Wesley, 1988.

[8]   Cutkosky, M. and Tenenbaum, M., "A Methodology and Computational Framwork for Concurrent Product and Process Design," 1989[To appear in Mechanism and Machine Theory, 1990].

[9]   Davis, R., Smith, R., "Negotiation as a Metaphor for Distributed Problem Solving," *Artificial Intelligence*, Vol. 20, pp. pg 63-109, 1983.

[10]   de Kleer, J., "Choices Without Backtracking," *Proceedings of the 4th NCAI*, Texas, AAAI, August 1984.

[11]   Doyle, J., "A Truth Maintenance System," *Artificial Intelligence*, Vol. 12, pp. 231-272, 1979.

[12]   Eastman, C. M., "Database Facilities for Engineering Design," *Proceedings of the IEEE*, Vol. 69, No. 10, pp. 1249-1263, October 1981.

[13]   Erman, L. D., Hayes-Roth, F., Lesser, V. R., and Reddy, D. R., "The Hearsay-II Speach Understanding System: Integrating Knowledge to Resolve Uncertainty," *Computing Surveys*, Vol. 12, No. 2, pp. 213-253, 1980.

[14]   Fenves, S., Flemming, U., Hendrickson, C., Maher, M., Schmitt, G., "An Integrated Software Environment for Building Design and Construction," *Computing in Civil Engineering: Microcomputers to Supercomputers*, 1988.

[15]   Fikes, R. and Kehler, T., "The Role of Frame-based Representation in Reasoning," *Communications of the ACM*, pp. 904-920, September 1988.

[16]   Groleau, N., *A Blackboard Architecture for Communication And Coordination in Design*, unpublished Master's Thesis, Dept. of Civil Engineering, M.I.T., Cambridge, MA 02139, 1989.

[17]   Hayes-Roth, B., "A Blackboard Architecture for Control," *Artificial Intelligence*, Vol. 26, No. 3, pp. 251-321, 1985, [Also Technical Report No. HPP 83-38, Stanford University.].

[18]   Howard, H.C., *Integrating Knowledge-Based Systems with Database Management Systems for Structural Engineering Applications*, unpublished Ph.D. Dissertation, Department of Civil Engineering, Carnegie Mellon University, Pittsburgh, PA 15213, 1986, [See also Special Issue of CAD, November 1985.].

[19]   Katz, R., *Information Management for Engineering Design*, Springer-Verlag , 1985.

[20]   "KEE Software Development Manual," Intellicorp, Palo Alto, 1987.

[21]   Klein, M., Lu, L., "Run-Time Conflict Resolution in Cooperative Design," *Workshop on AI in Engineering Design, AAAI-88*, 1988.

[22]   Krasner, H., "CSCW'86 Conference Summary Report," *AI Magazine*, pp. 87-88, Fall 1987.

[23]   Lai, K., Malone, T., Yu, K., "Object Lens: A Spreadsheet for Coopera-
        tive Work," *ACM Transactions on Office Information Systems,* Vol. 6,
        No. 4, pp. 332-353, October 1988.

[24]   Lander, S., Lesser, V., "Negotiation of Conflicts Among design Ex-
        perts," *Workshop on AI in Engineering Design, AAAI-88,* 1988.

[25]   Lochovsky, F., Ed., *Transactions on Office Information Systems: Special
        Issue on Object Oriented Databases, January,* ACM, 1987.

[26]   Maier, D., Stein, J., Otis, A., and Purdy, A., "Development of an Object-
        Oriented DBMS," *OOPSLA 86,* , pp. , 1986.

[27]   Marshall, R. D., et al., *Investigation of the Kansas City Hyatt Regency
        Walkways Collapse,* Technical Report Science Series 143, National
        Bureau of Standards, Washintong, D. C., May 1982.

[28]   Nii, P. and Brown, H., "Blackboard Architectures," AAAI Tutorial
        Notes No. HA2, 1987.

[29]   Nii, P., Fiegenbaum, E., Anton, J. and Rockmore, A., "Signal-to-Symbol
        Transformation: HASP/SIAP Case Study," *The AI Magazine,* Vol. 3,
        No. 2, pp. 23-35, Spring 1982.

[30]   Papadimitriou, C., *The Theory of Database Concurrency Control,* Com-
        puter Science Press, Inc., 1803 Research Boulevard, Rockville,
        Maryland 20850, 1986.

[31]   Pruitt, D., *Negotiation Behavior,* Academic Press, 1981.

[32]   Reddy, R., Erkes, J., Wood, R., "An Overview of DICE Architecture,"
        DARPA draft proposal by West Virginia University, 1988.

[33]   Rehak, D.R., Howard. H.C., and Sriram, D., "Architecture of an In-
        tegrated Knowledge Based Environment for Structural Engineering Ap-
        plications," in *Knowledge Engineering in Computer-Aided Design,*
        Gero, J., Ed., North Holland, 1985.

[34]   Rosenschein, J., Genesereth, M., "Deals Among Rational Agents,"
        Readings in Distributed Artificial Intelligence, pp. pg 227-234, 1988.

[35]   Sathi, A., *Cooperation Through Constraint Directed Negotiation: Study
        of Resource Reallocation Problems,* Technical Report, Carnegie-Mellon
        University, Robotics Institute, 1988.

[36]   Sathi, A., Morton, T., and Roth, S., "Callisto: An Intelligent Project
        Management System," *AI Magazine,* Vol. , pp. 34-52, Winter 1986.

[37]   Sriram, D., "DESTINY: A Model for Integrated Structural Design," In-
        ternational Journal for AI in Engineering, October, 1986.

[38] Sriram, D., *Knowledge-Based Approaches for Structural Design,* CM Publications, UK, 1987.

[39] Sriram, D., Ed., *Computer Aided Engineering: The Knowledge Frontier,* IESL, Dept. of Civil Engineering, M.I.T., 1989.

[40] Stefik, M. and Bobrow, D., "Object-Oriented Programming: Themes and Variations," *The AI Magazine,* pp. 40-62, Winter 1985.

PART IX: THE STATE OF THE FIELD

Chapter 13
CREATING A SCIENTIFIC COMMUNITY AT THE INTERFACE BETWEEN ENGINEERING DESIGN AND AI: A WORKSHOP REPORT

David Steier

ABSTRACT

On January 13-14, 1990, a workshop was held to discuss the topic of creating a scientific community at the interface between engineering design and AI, in order to identify problems and methods in the area that would facilitate the transfer and reuse of results. This report summarizes the workshop and followup sessions and identifies major trends in the field. This workshop was supported by the Engineering Design Research Center, an NSF Engineering Research Center.

13.1. MOTIVATION

It has been over thirty years since a group of engineers at Westinghouse created a program to design generators and motors using heuristic methods [1]. The program produced commercial-quality designs, and may have been the world's first expert system. Since then, the subfield of artificial intelligence devoted to engineering design applications has blossomed, especially in the last decade. In order to form a genuine scientific community out of the group of researchers active in this area, it is reasonable to ask a hard question: What do we know?

On January 13-14, 1990, twenty-three researchers in the field participated in a workshop designed to produce at least the beginnings of an answer. The workshop, held in Pittsburgh, Pennsylvania, was organized by the Engineering

Reprinted by permission from *AI Magazine*,
Vol. 11, Number 4, Winter 1990.

Design Research Center (EDRC) of Carnegie Mellon University (CMU) and sponsored by the National Science Foundation. Based on the observation that the research at the interface between engineering design and artificial intelligence was not as cumulative as it could be, the goal of the workshop was to identify problems and methods in the area that would facilitate the transfer and reuse of results. The workshop itself was by invitation only to permit focussed interaction, but was followed by a public session on the afternoon of January 24, at which the discussions of the workshop were reported and five additional CMU EDRC members made presentations and coordinated discussions. This paper reports on both workshop and followup sessions at several levels of abstraction: first, short summaries of each session are given; secondly, major trends, as evidenced by the presentations and discussions, are identified; and, finally, some assessment of how the workshop as a whole fulfilled its original goals.

The workshop organizers began by making their own assumptions explicit in the invitation to the participants:

1. It is both feasible and desirable to study engineering design scientifically. While gaps in current theories of engineering design exist, no portion of the design process is assumed *a priori* to be beyond analysis.

2. The design and implementation of computer programs that automate or support engineering design is a fruitful activity for researchers in the field.

3. Artificial intelligence concepts and techniques, including explicit representation of and reasoning about design goals, search of problem spaces, and methods for learning from experience, aid in the construction of engineering design systems.

4. It will be easier to make progress in this field if the research goals are structured so as to facilitate description, dissemination, and building on previous work.

We recognized that these assumptions would apply to only a subset of researchers interested in design (albeit a large subset), yet the belief was that some limiting set of assumptions had to be made to enable fruitful discussions. Part of the evidence that this tactic was successful was that relatively little time at the workshop and followup session was spent defining the term "design" itself.

With these assumptions made explicit, the organizers then took the unusual step of asking the prospective participants not to give detailed presentations of their own research. Participants were requested instead to prepare short (1-page) descriptions of two or three key results or techniques that they felt should be part of the core of a scientific discipline of engineering design and artificial intelligence. These descriptions were then copied and distributed to the participants prior to the workshop, and the workshop schedule was organized using clusters of related contributions as session themes (with some participants speaking multiple times). We should acknowledge that some of the invited researchers protested that the field was in fact not yet at a stage to identify such key results or techniques or that those results could not yet be called scientific. On the other hand, such objections did not deter these researchers from submitting a contribution or attending the workshop.

13.2. WORKSHOP AND FOLLOW-UP SESSIONS

By the time of the workshop, over two dozen contributions were submitted by the participants, and four session themes were chosen: models of design processes, coping with design complexity, representations for design, and goals and methods of the field. Each session consisted of five to ten individual presentations, interleaved with time for discussion. Each individual presentation was limited to eight minutes, followed by two minutes for clarification questions; time limits were strictly enforced with the aid of the VOID (Verbal Overrun Inhibitory Device), a darkroom timer hooked up to shut off the overhead projector at the end of a speaker's allotted time. Workshop participants soon demonstrated that they were capable of adapting to real-time constraints, and the VOID proved effective at keeping the workshop on schedule.

The session on models of design processes was divided into three topic areas, the first of which was using design process models to understand design. I began the session by presenting (on behalf of Allen Newell of Carnegie Mellon University) three lessons from human design: first, that human design appeared to be a combination of rapid kernel idea selection followed by a longer phase of elaboration and refinement of the kernel idea; second, that human design proceeds by alternately stipulating and retrieving partial functions and structures; and third, that specific structure plays an important role in the discovery of additional constraints for the design, in addition to the more commonly perceived role of approximation of the final design by a succession of partial artifacts. The next speaker, Mary Lou Maher (also of CMU), argued that three distinct models of design have been developed, and ought to be generally applicable to a variety of design applications, while being specific enough to be useful: decomposition of complex problems into simpler ones; case-based reasoning to draw on design experience in the form of specific episodes; and systematic transformations for design as in grammar-based approaches. Tim

Smithers of the University of Edinburgh noted that design ought to be considered as exploration rather than search of a single space with well-defined boundaries. The very ill-structured nature of most design means that the designer must spend a substantial portion of the design time refining the formulation of the problem. This claim was given additional support by John Gero of the University of Sydney, who gave further structure to the ill-structured problem of defining creative design with the figure below:

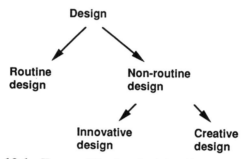

Figure 13-1: Types of Design Activity (According to Gero)

Routine design, Gero claimed, occurs when all functions and structures are known before design begins (thus parametric design is a special case of routine design). When the possible design structures are not known, design becomes innovative, but when neither structures nor the allowable design functions are known, then design becomes creative. Creative design extends the space of possible artifacts that the designer can produce.

The second topic area of the session was architectural support for design models. Barbara Hayes-Roth of Stanford University began by presenting a taxonomy of design problems, methods, and domains and gave instances of research to fit in the categories. Different design methods and heuristics are effective in different domains, depending on knowledge availability. She conjectured that a general and flexible computational architecture is best equipped to adapt to each situation, supporting the integration and acquisition of knowledge from a variety of sources. This conjecture is supported by the work of her and her colleagues on the BB1 blackboard system. B. Chandrasekaran of Ohio State University presented a slightly different approach towards the use of architectures to support design, focusing on task structures as a technique for specializing an architecture to the demands of a specific design method, such as propose-critique-modify. Hayes-Roth then presented a specific technique, based on a model of cross-domain analogy and Anderson's spreading activation memory in ACT*, for retrieving information within a successor to BB1. Chandrasekaran concluded the session by discussing the use of a functional representation of devices for retrieval of relevant design cases.

The third topic in the session discussed models of team design. Art Westerberg of CMU observed that most failures in design occur for organizational rather than technical reasons and introduced the idea that a design environment should provide support of the integration of many partial models of the artifact being designed. Sarosh Talukdar, also of CMU, argued that we need to devote more effort to the prediction and resolution of design conflicts, offering genetic algorithms as a possibility for capturing the evolutionary nature of the design process. John Goldak of Carleton suggested in his talk that one way to achieve success in supporting large-scale designs was to study carefully the way that VLSI design software has structured domain knowledge, and to transfer these methods to other applications. This suggestion was reinforced by John Hopcroft of Cornell, who moderated the discussion that followed.

The topic of the next session, coping with design complexity, was in one sense an extension of the session on design models. As our models represent the actual design process with increasing fidelity, they will become increasingly complex, and strategies will be needed to cope with that complexity. I began the session on the topic of progressive deepening. This pattern of repeated exploration of design paths in order to acquire new information on each pass is observed in human performance on many design, and design-related, tasks. Chris Tong of Rutgers University discussed how progressive deepening has been used in several automatic design systems (including FAD and DONTE). These use exploratory design (Volume II, Chapter 9) to add structure to a problem, rendering it amenable to methods of routine design. Tong also began the next topic: using knowledge from previous experience to reduce search. His SCALE algorithm improves the decomposition of a problem by reducing subproblem interactions. Jack Mostow, also of Rutgers, presented several lessons he and his colleagues have learned about the applications of derivational analogy to design; for example, design plans can be acquired by first recording user-selected transformation steps in an interactive design system, and then generalizing those steps using explanation-based learning techniques. The third topic of the session focussed on the use of abstraction. Ken MacCullum of the University of Strathclyde argued that the acquisition and use of abstractions at multiple levels is crucial for good designs. Mostow then discussed automated techniques for discovering useful abstractions, illustrating them with systems, ABSOLVER and POLLYANNA, built by his students. The session was concluded by several presentations by Mostow, Tong, and Tom Dietterich (of Oregon State University) on the use of knowledge compilation. This term describes a class of techniques for producing efficient search algorithms for specific classes of problems, generated by integrating algorithm design and domain-specific knowledge. [See Section 1.5.3.3].

The third session concentrated on representations for design. The better adapted a representation is to a particular design problem, the easier it will be for a designer to search the space of designs for a high-quality solution. Chuck

Eastman of UCLA discussed two representational concepts based on database techniques: first, data modeling methods for engineering product models, and second, design transactions, in which each transaction satisfies certain integrity constraints. He also discussed shape grammars, a formalism for representing geometric knowledge pioneered by George Stiny. Gero then summarized the motivations behind the concept of design prototypes, which independently represent the function, structure and behavior of engineered artifacts. Talukdar discussed the representation of designs as n-tuples of partial models. Tetsuo Tomiyama of Tokyo had planned on discussing the role of logical formalisms and qualitative physics for representing engineering knowledge, but was not able to attend the workshop. Sanjay Mittal of Xerox Palo Alto Research Center discussed dynamic constraint satisfaction for engineering design. Panos Papalambros of the University of Michigan presented an example of how structural design problems could be solved using mathematical optimization, and argued that methods for combining AI and optimization techniques ought to be given more attention in research on engineering design.

The final session took a broader perspective and directly addressed the nature of a scientific community in this area. Herb Simon of CMU initiated the topic on theories and data in design science by presenting some of his ideas on problem formulation, satisficing, the order-dependence of design, and the need to combine techniques from whatever disciplines are appropriate. This talk was followed by Steve Fenves' (also of CMU) talk on how design theories should serve to explain design processes. John Gero emphasized how design science must work more strongly to collect data on design. In the second topic of this session, the topic suggested by Gero was explored in detail, in terms of what are the sources of data on design. Dietterich discussed the use of protocol analysis, based on his studies of mechanical design, while I discussed the need to perform comparative design studies using common examples, drawing on my work on algorithm design studies. The session and the workshop was concluded by a discussion on infrastructure for the field.

The followup session, held two weeks later, began with a brief review of the workshop presentations and discussions, followed by an extension of the format to allow additional researchers to present their "golden nuggets" for design science, some of which were not explored in depth during the workshop itself. All the speakers were affiliated with CMU. Newell presented his conception of how the field should progress in terms of the diagram below:

AI researchers produce systems that design, each of which serves as a data point from which to induce a science of design, since each system is a detailed design model. Studies of human design are used to guide the creation of design systems, since humans are currently the only intelligent agents possessing the robustness and flexibility that we would like our design systems to have.

Mark Fox listed several examples of design process, such as indexing, filtering, analogy, composition, and constraint satisfaction, and discussed the role of opportunism in design. Ulrich Flemming

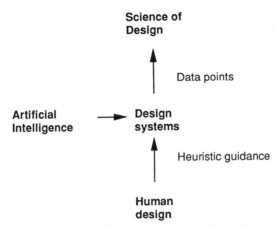

Figure 13-2: Evolution of Design Systems (According to Newell)

argued that we should focus on formal models of design to produce designers that did not have the weaknesses of humans, dividing the labor so that human capabilities would only be relied upon for guidance as needed. Rob Woodbury extended this further with specific examples of shape grammars, structure and solid grammars, and non-manifold representations. Such representations, he noted, had desirable properties in terms of expressiveness and simplicity, and enabled efficient search in geometric-based domains such as building and truss design. Susan Finger argued that we should be focusing more on representation of the designed artifacts themselves than on the design process, claiming that process-oriented research hasn't made as much progress as artifact-oriented research, such as the development of formal geometric models. Dan Siewiorek presented several nuggets from his work with his student on building knowledge-based design systems, including the observations that engineers developing heuristics for different synthesis problems are already actually practicing AI without knowing it, and that in his experience it was more productive (in terms of knowledge integrated into a system per unit time) to train domain experts in AI techniques than vice versa. Ultimately though, automated, domain specific knowledge acquisition systems will be required to achieve truly expert performance.

13.3. MAJOR TRENDS

I noted five trends that underlay much of the discussion at the workshop and followup session:

1. A near-consensus on the complementary roles of function and structure in design: Terms such as "function" and "structure" were often used in different senses by the workshop participants, and some attention was devoted to clarification of these. Some used the term "function" in the sense of the purpose of some structure relative to the design goals, where others used function to describe the behavior exhibited by some structure. Newell's abstract for the workshop noted that descriptions of structures are generally much less problematical than the description of functions (in the first sense), which are conceptual entities used to decompose the design process. In either case it does not seem that design is a straightforward mapping from functional specifications to structural descriptions. Both structure and function seem to play important roles in retrieving design knowledge (as noted by Newell, Chandrasekaran, and others) and both are elaborated during design. Yet we are far from a complete theory of functions, or of a theory to link function and structure. One concept, which did not come up in the abstracts[2] or discussions, but which does seem to provide a function-structure link, is that of design features: features are aspects of the structure of a design, but their importance for any given process depends on the functions being fulfilled.

2. The beginnings of a common vocabulary for modeling design: As noted by Hayes-Roth, a theory of design must describe problem types, methods for problem solutions, and problem domains. She proposed "arrangement" as a candidate problem type, which she found occurring in domains as diverse as biochemistry, construction management, and architectural layout. Examples of methods are "propose-critique-modify" (Chandrasekaran), "decomposition" (Maher), and "progressive deepening" (Steier, Tong). A variety of design domains were discussed at the workshop; among them were aircraft design (Goldak), digital logic (Mostow, Tong), and structural design (Dietterich, Papalambros). Terms to describe approaches for acquiring design data, such as protocol analysis (Dietterich), and for implementing design models, such as design architectures (Hayes-Roth, Chandrasekaran) were also used at the workshop. As Fenves pointed out, the establishment of such common vocabulary is one of the prerequisites for the establishment of a scientific community in this area. Yet, the field still seems to have some distance to go before

---

[2]As pointed out by Caroline Hayes, who has built a feature-based machining planner.

terms are defined with precision, and commonly used in comparison and evaluation of work.

3. An increasing recognition of the importance of studying design-in-the-large: Several speakers (Westerberg, Goldak, Hopcroft, Talukdar) noted that issues arise in team design that are not often encountered in research on individual designers. Among these are the knowledge representation and integration process for multiple partial artifact models, and the prediction and resolution of design conflicts. Tools, such as n-Dim (Westerberg), that recognize the organizational context in which most real design takes place are just beginning to be built, but there was a clear sense in the workshop that the issues ought to be studied further, and that there is a need to devote more research effort to observing and analyzing large-scale design.

4. A growing collection of methods for acquiring and applying abstract or approximate design knowledge: Design prototypes (Gero) are one method for capturing general design knowledge. Abstract design algorithms (Mostow, Tong, Dietterich), and design task structures (Chandrasekaran) are others. A notable development is the use of knowledge compilation to specialize abstract algorithm schemes into efficient algorithms in several design domains. MacCullum observed that an important issue in developing systems that use such abstract knowledge is how the abstractions are acquired. Mittal observed from his studies of designers in the PRIDE (Volume I, Chapter 9) and COSSACK projects that representations used to articulate knowledge may not be the most appropriate for use in automated design systems. A wide range of methods for producing abstractions were discussed. Some of the methods assigned a central role to "knowledge engineering," suggesting systematic, formal analysis of the artifacts (Flemming, Woodbury, Finger), comparative study of common examples (Steier), or protocol analysis (Dietterich). Others have stressed automated techniques for acquiring such knowledge from experience (Mostow, Tong). Mostow in particular has recently focused on methods for acquiring useful approximations. It is likely that robust design systems will result from combinations of these methods (Simon).

5. The continuing proliferation of radically different knowledge representations for design: Several participants (Mittal, Newell, Simon, Smithers, Sriram) noted that both constraint formulation and satisfaction are important in design. Mittal's research has focused on representing such constraints and their dynamic nature explicitly. Other representations mentioned at the workshop and follow-up sessions included shape grammars (Eastman, Flemming, Woodbury), engineering data models (Eastman) or n-tuples of such models (Talukdar), logical expressions (Tomiyama), systems of equations (Papalambros P.) and bond graphs (Finger). All the participants seemed to agree that a representation should be adapted to the application for a special-purpose system. But the development of a theory that would guide the selection of representations for a particular application, especially a complex one, remains a research question. One way to acquire the data on which to build such a theory would be to encode the same engineering design knowledge in different representations for the purposes of evaluating the efficacy of different representations in a particular context.

13.4. GOING FOR THE GOLD

When we began to plan for the workshop, we had intended for this report to include a list of two dozen or so golden nuggets that would form a core of a scientific discipline of engineering design and AI. While the workshop was successful in terms of generating stimulating presentations and discussion, there was not very much progress towards the production of the list, nor is there a clear method -- at least not to this participant -- for producing such a list by post-workshop armchair analysis. The workshop did produce some candidates for the list, although not the two dozen we had hoped for. These candidates would include techniques such as the use of functions as memory retrieval cues, and terms such as design task structures or derivational analogy. But in retrospect it is clear that any field which claims to have golden nuggets must also propose the assay for those nuggets. And in fact methods for evaluating contributions were absent from most of the workshop presentations.

I am still convinced that a list of golden nuggets for this field (eventually leading to a handbook of techniques) would be both very useful to have and feasible to produce, but perhaps only with additional research and several more iterations on the workshop format. We probably need to understand better what is meant by a theory of design, a process begun over twenty years ago by Herb Simon's book [2]. We need to concentrate on the role of data in such theories, perhaps establishing repositories for design case histories, as was suggested at the workshop. Both design processes and designed artifacts need to be studied. And we would need further convergence on assumptions, vocabulary and certainly, criteria for evaluating contributions. I believe the workshop provided a useful forum to discuss these issues, essential for further progress of the budding scientific community in engineering design and AI.

13.5. ACKNOWLEDGEMENTS

The workshop was sponsored by NSF grant ECD-8943164. I am grateful to Professors Allen Newell, Art Westerberg, and Mary Lou Maher for assistance in organizing the workshop, and to the head of the EDRC, Fritz Prinz, and EDRC staff Nancy Monda, Mary L. Ray, and Sylvia Walters for their administrative support. Newell, Westerberg, and Peter Patel-Schneider also provided useful comments on an earlier draft of this report.

13.6. BIBLIOGRAPHY

[1]　Godwin, G. L., "Digital Computers Tap out Designs for Large Motors...Fast," *Power,* April 1958.

[2]　Simon, H. A., *The Sciences of the Artificial,* MIT Press, Cambridge, MA, 1969.

Index

Abstraction 34, 373, 377
  part-whole 15, 282
  product 292
Abstraction levels 16, 30, 258
  gap in 18, 20
ACT* 372
ADAM 31
Adaptation and reuse of design cases 27
  replay of a design plan 28
ADS 242
AIR-CYL 23
Aircraft engine turbine design 247
ALADIN 29, 30
Algorithm level of system description 7, 46
Analysis 9, 264, 307
  codes 181, 235, 238, 251
  finite element 30, 35, 105, 106
  models for 65, 333
  of criticality 32
  of stress 35
  See also evaluation
AND/OR graph
  design space as 16
  search of 22
ANVIL-5000^{TM} 87
Architecture, design process 255, 258,
  278, 372
  blackboard 10, 44, 267, 291, 295, 311,
  316, 321
ARGO 27, 37
Artifact, designed 3
Artificial Intelligence 5
  and Computer-Aided Design 12
  as a software engineering methodology 8
Assembly process 225
Assembly, design for 219
ATMS 8, 46
AutoCAD^{TM} 149, 172
Availability of a designed artifact 279
Axiomatic design 264
Axworthy, A. 125

Backtracking
  chronological 24

dependency-directed 24
knowledge-directed 24
See also belief revision
BB1 372
Behavior of an artifact 3
Belief revision 6
  ATMS 8, 46
  See also backtracking
BIOSEP 24
Blackboard architecture 10, 44, 267, 291, 295,
  311, 316, 321
Bobrow, D. 310
BOGART 27, 37
Boiler plant design 127
Boothroyd, G. 225
Brown, D. 292
Budgeting 4, 31, 32
BUILDER 355
Building a knowledge-based design tool 35
Business Week 42

C 181
C++ 38, 45
CADET 27
CALLISTO 294, 318
Carbonell, J. 340
Case database 26, 34, 267
  See also part library, knowledge base
Case-based reasoning 12, 26, 27, 37
  adaptation and reuse of design cases 27
  representation/organization of design cases
  27
  retrieval of design cases 27
  storage of design cases 27
Causal knowledge 333
Chandrasekaran, B. 292, 372, 376, 377
Chandrasekaren, B. 376
Chao, N. 199
Cherneff, J. 303
CHIPPE 24
Chronological backtracking 24
Classification task: why design isn't one 3
Classifying a design task 18, 37

Cognition MCAE^{TM} 15
Cognitive models of design 371
    See also protocol analysis
COLAB 314
Cold forging 105
Collaborative product development 43
Commercially available design tools 38
    AutoCAD^{TM} 149, 172
    Cognition MCAE^{TM} 15
    Concept Modeller^{TM} 39, 88, 179, 180
    Design^{++tm} 39, 125
    DesignView^{TM} 15, 39, 110
    ICAD Browser 88
    ICAD^{TM} 39, 81, 88
    NEXPERT^{TM} 39, 112
    vendors for 49
    VP-EXPERT^{TM} 46
    See also design tools
Common sense knowledge 2, 32
    See also qualitative reasoning
Compiler
    of knowledge 38, 39
    See also knowledge compiler
Complexity
    exponential design time 24
    of subproblem interactions 18
    polynomial design time 24
Composite objects 359
Compositional interactions 21
Computer-Aided Design (CAD) 9, 12
    and Artificial Intelligence 12
    knowledge-based 199
Computer-integrated manufacturing 255, 256
Concept Modeller^{TM} 39, 88, 179, 180
Concurrency control 361
Concurrent engineering 43, 103, 111, 114, 164
Configuration spaces 35
Configuration task 20
Consistency maintenance
    using a dependency network 170
    See also belief revision, constraint
        processing
Constraint graph 15
Constraint processing 6, 22
    constraint propagation 39
    dynamic constraint satisfaction 374
    See also what if analysis
Constraint satisfaction problem (CSP) 22
Constraint, design 21, 265
    functional 179
    geometric 64, 185
    global 21, 23
    local 21, 332
    manufacturing 179
    non-directed 110
    nonlinear equation 111

product cost 64
resource limitation 4
semi-local 21, 332
service 64
Construction planning 306, 308
Contract net 319
Control heuristic 12
Control knowledge 19
Control strategy for a design process 6, 10
    hillclimbing in the space of problem
        formulations 31
    least commitment 30
    multi-layered 30
    planning 30
    progressive deepening 373
    top-down refinement of design plans 31
Conventional design methods
    AND/OR graph search 22
    constraint satisfaction 22
    integer programming 22
    linear programming 22
    multi-objective optimization techniques 22
Conventional routine design 22
Cooling fan design 250
Cooperative engineering design 43, 277, 303
    coordination of 210, 313
Cornett, D. 235
Correctness of a design 3
COSSACK 377
Cost model (for estimating costs) 226
Cost, product 256
Creative design 2
    See also design process model
Creative design task 19
Criticality analysis 32
Customizable knowledge base 223
CYC 32

DARPA DICE 43, 314
Database management system 181, 205, 290,
    296, 312, 317, 361
Database, relational 90
    INGRES 205
    Oracle 90, 150, 205
    Sybase 150
Davis, R. 319
Decision-making, design 264, 269
Deep knowledge 333
Dependency-directed backtracking 24
Derivation problems 306
Derivation-Formation spectrum 306
Design 259, 267, 277, 278, 306
    as part of a larger engineering process 41
Design axioms 264
Design capture 9
Design exploration 31

Design for assembly 219
Design for manufacturability 2, 103, 211, 219,
    226, 256, 257, 259, 278, 306
Design for Manufacturability auditing system
    211
Design information
    See also information, design
Design of
    aircraft engine turbines 247
    aluminum alloys 29
    analog circuits 34
    architectures 362
    automobile transmissions 15
    ball bearings 63
    boiler plants 127
    cold forged products 106
    cold forging forming sequences 106
    cooling fans 250
    document transport subsystems 185
    fuser rolls 190
    gear chains 39
    generators 369
    house floorplans 39
    injection molds 82
    light-weight load-bearing structures 34
    mechanical linkages 34
    molecular electronic structures 248
    motors 250, 369
    power supplies 250
    scan subsystems in copiers 185
    software 2
    spatial layouts 34
    turbine blades, 3D 251
    turbine disks for aircraft engineers 278
    VLSI chips 9, 16, 17, 20, 21, 31, 39
Design plan 26, 31
Design process 1
Design process model 18
    conceptual design 145, 236, 307
    creative design 2
    detailed design 145, 236, 308
    innovative design 26
    preliminary design 145, 236
    routine design 22
Design process operation
    adaptation and reuse of design cases 27
    constraint processing 22
    decomposition 16
    identifying key design parameters 235
    implementation 16, 22
    optimization 16, 22
    patching 16, 22
    planning the design process 30
    problem re-structuring 24
    refinement 15, 16, 22
    retrieval of design cases 27

    storage of design cases 27
    structural mutation of a design 28
Design review, global 308
Design rules 155
Design space 14
Design task 1, 18
    classification of a 18, 37, 47
    dimensions for classifying a 18
    See also task, design
Design tools
    commercially available 38
    knowledge-based 3
    See also commercially available design tools
Design^{++tm} 39, 125
Designed artifact 3
DESIGNER 24
DesignView^{TM} 15, 38, 39, 110
DESTINY 316
DFMA 43
DFX (Design for Manufacturability, Testability,
    etc.) 168, 199
    auditors 202
    Design for availability 279
    design for maintainability 168
    Design for manufacturability 2, 103, 168,
        211, 219, 226, 256, 257, 259, 278, 306
    Design for reliability 279
    Design for serviceability 279
    design for testability 168
    design for transportability 168
DICE (MIT) 303
Die casting 232
Diebenkorn, R. 293
Dietterich, T. 373, 374, 376, 377
DIOGENES 316
Directed interviewing (to acquire design
    knowledge) 58
DMA 44
Document transport subsystem design 185
Documentation, computer-aided 218
Domain theory, design 36
Domain-independent shell 39
    EMYCIN 65
    Engineous 43, 235
DONTE 32, 373
Dual design partner paradigm 282
DXF files 111, 121, 149

Eastman, C. 373, 377
ECMG 43
ELF 39
Emacs 161
EMYCIN 65
Engineering
    over the wall 42, 168
Engineering design 1, 3

as a science 370
Engineous 43, 235
Epistemological adequacy 8
Ericsson, K. 58
ESKIMO 82
Estimation of costs 223
Evaluation 118, 307
    fitness profile 284
    of design plans 31
    of machining cost 225
    See also analysis
EXFAN 250
Expert Cost and Manufacturability Guide 224
Expert systems paradigm 5
Exploratory design 31, 372, 373
    See also control strategy for a design
        process

Factory of the future 256, 255, 258, 278
FAD 373
Features, design 376
Fenves, S. 315, 374, 376
Finger, S. 375, 377
Finite element analysis 30, 35, 105, 106
Fitness profile 284
Flemming, U. 374, 377
Forging process 225
Formation problems 306
FORMEX 114
FORTRAN 131, 181, 226, 258
Fox, M. 374
Frame-based language 6
    See also object-oriented programming
FrameMaker 173
Function of the artifact 3
Function-structure mapping 3, 371, 376
Functional requirements 257, 265, 269
Functionality-preserving transformation 28
Fuser roll design 190

Generative design knowledge 19
Generator design 369
Genetic algorithms 235
    See also learning in design
Geometric models 375
Geometric reasoning 2, 48, 203, 207
    analysis 35
    simulation 35
    synthesis 34
    variational geometry 110
Geometric representations 361
    3D 97
Gero, J. 372, 374, 377
Global constraint 21, 23
Global objective function 24
Goldak, J. 373, 376, 377

Graph
    constraint 15
Groleau, M. 303

Hankins, B. 223
Hayes, C. 376
Hayes-Roth, B. 372, 376
Heatley, L. 179
Heuristic adequacy 8
Heuristic knowledge 264, 333
Hierarchical refinement 39
Hierarchy, part 89, 258
Hillclimbing, knowledge-directed 24
Hopcroft, J. 373, 377
Human design 371

IBDE 315
IBPD 125, 127
ICAD^{TM} 39, 81, 88, 90
IDES 71
IDL 88, 89, 90
IGES (Initial Graphics Exchange Specification)
    files 89, 92, 111, 121
Ill-structured problem 8, 46
Implementation (of a design specification) 15, 22
Independence of functional requirements 265
Information content, minimization of design 265
INGRES database 205
Injection mold design 82, 83, 98
Innovative design 2
    correlated with incomplete or incorrect
        design knowledge 47
    via genetic algorithms 235
    vs. conservative design 281, 282
Innovative design task 19, 47
Instantiation (of a generic design part) 90
Integer programming 22
Integrated product development 43
Integration of
    conceptual, preliminary, and detailed design
        237
    design and manufacturing 44, 2, 103, 211,
        219, 226, 256, 257, 259, 278, 306
    product and process design 111, 278
Interactions
    between parts 4
    complexity of subproblem 18
    compositional 21
    functional 21
    local 5
    physical 21
    represented as constraints 21
    resource 21
    strong 4, 21
    subproblem 4
    weak 5, 21

Interleaf 173
Interval arithmetic 15
Invention, interaction-based 34
    See also innovative design, creative design
Inverting the structure-function mapping 27

Joskowicz, L. 35

KADBASE 316
Karras, T. 125
Katajamki, M. 125
KBEmacs 294
KBSDE 39
KEE^{TM} 38, 46, 125, 146, 149, 355
Kernel idea for a design 371
Kim, S. 255
Kinematics
    kinematic equivalence 35
    kinematic simulation 35
    kinematic synthesis 34
Knight, W. 225
Knowledge acquisition 2, 36
    tailored for a design process model 36
    via decision flow charts 57
    via directed interviewing 58, 61
Knowledge base
    customizable 223
    of analysis codes 240
    of cold forging operations 115
    of design parameters 240
    of design rules 240
    of meta-knowledge 288
    of process knowledge 288, 293
    of product knowledge 288, 291
Knowledge compiler 38, 39, 373
    DIOGENES 39
    ELF 39
    KBSDE 39
    WRIGHT 39
Knowledge engineering 58
Knowledge level of system description 7, 46
Knowledge module 323
    Critic 323, 334
    Quantitative 323, 334
    Specialist 323, 333
    Strategy 323, 333
Knowledge sources 323
    combining multiple 29
Knowledge-based design tools 3, 186, 203, 207,
    311
Knowledge-based paradigm 7
Knowledge-based routine design 22

Lander, S. 320
Layout level for VLSI design 20
Learning in design 12, 243, 373

Least commitment strategy 30
    See also control strategy for a design process
Lehtimaki, H. 125
Lesser, V. 320
Levels for describing knowledge-based systems
    algorithm level 7
    knowledge level 7
    program level 7
Levels of abstraction 16, 30, 258
Levitt, R. 125
Linear interpolation 31
Linear programming 22
LISP 6, 38, 65, 157, 169, 186, 192
Local constraint 21
Locking attribute values 170
Logcher, L. 303
Logic level for VLSI design 20
London, P. 223
Luby, S. 223, 225

MacCullum, K. 373, 377
MacDraw 64, 74
Macro-decision formation 32
Macrorules 27
Maher, M. 371, 376
Manufacturability, design for 2, 103, 211, 219,
    226, 256, 257, 259, 278, 306
Marketing, product 41, 277
Mechanical design 9
Mechanical engineer 227
Metal design 227
Metaplanning
    See also planning the design process
Microplanner 6
Microprocessor 269
Minsky, M. 61
Mishelevich, D. 125
Missing design knowledge, compensating for 26
MIT-DICE 44
Mitchell, T. 286
Mittal, S. 374, 377
Mixins 89
Model-based reasoning 267
Modularity 257
Mold design 81
Molecular electronic structure design 248
MOLGEN 30
MOSAIC 34
Mostow, J. 286, 373, 376, 377
Motif 183
Motor design 250, 369
Multi-layered control strategy 30
    See also control strategy for a design process
Multi-objective optimization techniques 22
Multiple knowledge sources, combining 29
Multiple objectives, satisfying 23, 30

Mutation of a design, structural 26, 28

N-Dim 377
Navinchandra, D. 293
Negotiation 313
  Computer-aided 318
  Multistage 319
  Pruitt 313
Newell, A. 371, 374, 376
NEXPERT™ 38, 112
NEXT-CUT 315
Non-uniform rational B-spline 181
Nonlinear equations 111

Object-oriented database 17
  ENCORE 316
  GEMSTONE 316
  ORION/ITASCA 316
Object-oriented programming 6, 10, 88, 146, 310
Object-oriented semantic network 296
Objectives, multiple 23
OPAL/GEMSTONE 45
Open architecture 146, 203, 210
OPS5 6, 36, 38, 65, 250
Optimization 16, 23, 235, 241, 374
  combined with AI techniques 235, 374
Optimization criteria 4
Optimization rule 18
Oracle database 90, 150, 205
Organization of design cases 27
Over the wall engineering 42, 168

Papalambros, P. 374, 376, 377
Parameter instantiation task 20, 22
  See also design task
Parameters, design 265, 276
  identifying key 235
Parametric modeler 38
Pareto-optimal solution 23
PARMENIDES/FRULEKIT 340
Part hierarchy 89, 258
Part library 155
  geometric parts 181
  See also case database, knowledge base
Part-whole abstraction 282
Pascal 258
Patching 16, 23
  innovative 18
  patching rule 18
PDES 296
Phillips, J. 292
Physical interactions 21
Physical system 33
Plan, design 26, 31
  See also design plan
Planning the design process 30
  See also control strategy for a design process

Poli, C. 225
Powell, D. 235
Power supply design 250
Preserving functionality 28
PRIDE 24, 377
Problem identification 307
Problem re-structuring, knowledge-directed 24
Process, physical 111, 268, 278
Product 111, 268, 278
Product abstraction 292
Product database 288
Program level of system description 7, 46
Programming language or environment
  ANVIL-5000™ 87
  C 181
  C++ 38, 45
  FORTRAN 131, 181, 226, 258
  KEE™ 38, 46, 125, 146, 149
  LISP 6, 38, 157, 169, 186, 192
  Microplanner 6
  OPAL/GEMSTONE 45
  OPS5 6, 36, 38, 65, 250
  PARMENIDES/FRULEKIT 340
  Pascal 258
  Prolog 121, 265
Programming paradigm 6
  frame-based language 6
  object-oriented programming 6, 10, 88, 146, 310
  rule-based language 6, 155, 240
Progressive deepening 373
  See also control strategy for a design process, exploratory design
Prolog 121, 265
Protocol analysis 58, 374, 377
  See also cognitive models of design

Qualitative reasoning 29, 32
  See also common sense knowledge

Re-budgeting 32
Re-design 34
  See also what if analysis
Refinement 15, 22
Refinement rule 18
Reformulation of the design problem
  hierarchical 32
  knowledge-directed 24
Relational database 90, 149
Reliability of a designed artifact 64, 279
Replay of a design plan 28
Representation of
  design rules 155, 240
  designs 27, 38, 203
  designs using a semantic network 296
Resource interactions 21
Resource limitation 4

Retrieval of design cases 27
    See also case-based reasoning
Riitahuhta, A. 125
Robustness of the knowledge representation 193
Rough design 32
Routine design 2, 19, 30
    conventional 22
    hypothesis 31
    iterative, knowledge-based 23
    knowledge-based 22
    non-iterative, knowledge-based 23
    See also design process model
Routine design task 19, 22, 47, 81
Routing 39, 225
Rule-based language 6, 155, 240

SALT 36
Sapossnek, M. 223
Scalability of design methods and tools 180, 192
Scan subsystem design 185
Science, engineering design as a 370
Search algorithm 5, 7, 8
Search paradigm 5, 7
Search, design as 14
Semi-local constraint 21
Serrano, D. 15
Serviceability of a designed artifact 279
Sevenler, K. 105
Shape grammars 374
Shell, domain-independent design 39
    Engineous 43, 237
Shoenfeld, A. 58
Sieworek, D. 375
Signal processing 312
Silvestri, M. 281
Simon, H. 58, 374, 377, 378
Simulation 30
    qualitative kinematic 35
Simultaneous engineering 43
Smith, R. 319
Smithers, T. 371, 377
Software engineering 8
Solid modeling 200, 207
Space, design 14
Spear, W. 179
Specialist KMs 333
Specification generation 307
Speech recognition 312
SQL 149, 299
Sriram, D. 1, 303, 377
Steele, G. 6
Stefik, M. 30, 310
Steier, D. 369, 376, 377
Stiny, G. 374
Storage of design cases 27
    See also case-based reasoning

Strong interactions 21
Structure configuration task
    See also design task
Structure of the artifact 3
Structure synthesis task 20
    See also design task
Structure-sharing 4
Surface knowledge 333
Sybase 150
Symbolic referencing 89
Synthesis 9
Systematic progress, guaranteeing 10

Talukdar, S. 373, 377
Task, design
    decomposition of 22
    parameter instantiation task 20
    structure configuration task 20
    structure synthesis task 20
Team design 306, 372, 377
Technology, implementation 3
Textbook knowledge 333
Tomiyama, T. 286, 374, 377
Tong, C. 1, 373, 376, 377
Tong, S. 235
Toolbox, design 10
Tradeoff, design 23
Truth maintenance system 8, 46
Turbine design 251, 278

Unigraphics[TM] 87
User interface tools
    Emacs 161
    FrameMaker 173
    Interleaf 173
    MacDraw 64, 74
    Motif 183
    X Windows 176, 183

Variational geometry 110
Vendors of commercial design tools 49
Version management 361
Version, design 10
VEXED 23
Visualization 43, 58, 149, 312
VLSI design 9, 16, 17, 20, 21, 31, 39
VP-EXPERT[TM] 46
VT 24, 37

Waldron, K. 57
Waldron, M. 57
Waters, R. 287, 294
Weak interactions 21
Weak method 5
Well-structured problem 9
    See also routine design

Westerberg, A. 373, 377
What if analysis 186, 229
    via back-solving 110
        See also constraint processing, re-design
Wilensky, R. 30
Williams, B. 34
Woodbury, R. 375, 377
Workshop on AI in Design (1990) 369
WRIGHT 24, 34, 39
Wright, Frank Lloyd 293

X Windows 176, 183, 362